Patrick Moore's Practic

For other titles published in the series, go to
http://www.springer.com/series/3192

Starlight

An Introduction to Stellar Physics for Amateurs

Keith Robinson

 Springer

Keith Robinson
4 Bedford Place
Scotforth, Lancaster
United Kingdom
starlightskies@talktalk.net

ISSN 1431-9756
ISBN 978-1-4419-0707-3   e-ISBN 978-1-4419-0708-0
DOI 10.1007/978-1-4419-0708-0
Springer New York Dordrecht Heidelberg London

Library of Congress Control Number: 2009931800

Printed on acid-free paper

Springer is part of Springer Science+Business Media (www.springer.com)

Acknowledgements

My very grateful thanks go to Jean-Francois LeBorgne of the Laboratoire Astrophysique de Toulouse, for permission to reproduce spectra from their STELIB library. Many thanks also to Gary Billings, Jeff Hopkins, Robin Leadbeater, and Daniel Majaess of the American Association of Variable Stars Photometry Discussion Group for kindly providing information regarding online sources of spectra.

Introducing a little bit of mathematics into a book like this can be a risky business, so it's always good to have someone who can offer critical appraisal. My sincere thanks go to my great friend and amateur astronomer par excellence, Denis Buczynski, for doing this in respect of chapter *Starlight by Numbers* and offering some helpful criticisms and suggestions.

Many thanks to Maury Solomon, Harry Blom, and all at Springer, New York, for their help and enthusiastic support during the writing of the book and also to John Watson for his great enthusiasm and encouragement over the original idea for the book.

Finally my heartfelt gratitude goes to my wife Elizabeth for reading the manuscript and offering many helpful comments and suggestions.

All diagrams were prepared by me.

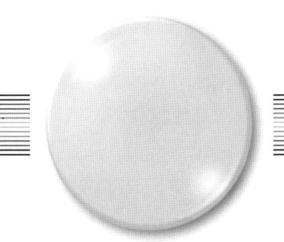

Contents

About the Author

Keith Robinson became interested in astronomy at the age of eight after seeing the TV documentary "Universe" produced by the National Film Board of Canada. He graduated with a degree in physics at the University of Lancaster (UK) in the early 1970s and subsequently worked as observatory superintendent at the Jeremiah Horrocks Observatory in the Centre for Astrophysics at the University of Central Lancashire. Here he obtained his PhD for research on modeling the Balmer emission line profiles in the spectra of symbiotic stars.

Resident in his native northwest of England and married, with one daughter, he is very much involved with his local astronomy club and works as a full-time writer and educator. He became a member of the American Association of Variable Star Observers in 2001.

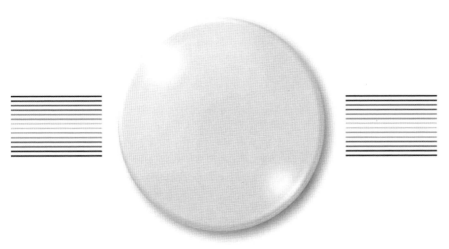

A River of Starlight

Back in the early 1980s, I swapped a pretty good quality 3 inch refractor for a "so so" $8\frac{1}{2}$ inch Newtonian reflector on a fairly rough and ready German equatorial mounting (to be fair it did have manual slow motion drives which, after a bit of practice, I got to be fairly good at using, and it did have quite large and easy to read setting circles). I guess I was greedy for that extra aperture, which would enable me to see deeper and fainter, and indeed I had a thoroughly enjoyable couple of years observing deep-sky objects.

The fact is, as a teenage amateur astronomer in the late 1960s (yes, I did manage somehow to find the time), I had read that once one had progressed beyond the beginner level, one should seriously consider specializing in some specific area of observational astronomy. To be honest I didn't really like the idea of, for example, spending the rest of my life just observing Jupiter (no disrespect to Jupiter observers). The trouble was that the books of the time didn't seem to make any mention of the fact that there was no "law" that said that if you specialized in one area of astronomy you were not allowed to investigate other areas. On the contrary, there was this sense that you were strongly encouraged to specialize in one thing. I did at the time rather like the idea of observing what were referred to as "nebulae and galaxies," etc. (I never came across the phrase "deep-sky object" until the 1970s.) However I remember someone – probably an older kid at my school – telling me that there was no useful work that could be done in this area by amateurs (try telling that to supernova hunters), so there was no point to it.

K. Robinson, *Starlight*, Patrick Moore's Practical Astronomy Series,
DOI 10.1007/978-1-4419-0708-0_1, © Springer Science+Business Media, LLC 2009

Now of course, deep-sky observing is quite rightly one of the most popular areas of amateur astronomy, whether or not it is scientifically useful; hence the acquisition of the Newtonian and "Yah! Boo! Sucks!" to that long forgotten school kid. However, one clear August evening things changed.

I had taken the trouble to polar align my equatorial mount, so that I could use the scope's setting circles, and I remember quite a feeling of satisfaction at being able to locate the Dumbbell Nebula, M27, without even looking at the sky. I also remember that upon enjoying the view of what I reckon is a more impressive planetary nebula than the Ring Nebula, M57, I felt a sense of wanting to do some kind of observing that involved more than just looking.

Astrophotography was out of the question with my scope, so it had to be some kind of visual observing. On looking up at the sky, I then noticed that Algol (the famous eclipsing variable star in Perseus) looked distinctly dimmer than its nearby neighbor Mirfak (Alpha Persei). I happened to have at hand the *Handbook of the British Astronomical Association*, and sure enough, Algol was around half an hour from minimum magnitude. This was the first time I had ever seen a variable star "in action."

Some years previously, when considering my "choices" for a specialized area of amateur astronomy, I had been distinctly put off the idea of observing variable stars, simply because those aforementioned books of the time seemed to suggest that, not only was it possible by making visual observations to estimate the magnitude of a variable star to an accuracy of one tenth of a magnitude, but that this was actually some kind of "standard" that was expected. Maybe I misinterpreted what I'd read, but one thing's for sure. I don't recall coming across any book that gave an illustration of a real (rather than a stylized) light curve of a variable star, which showed the obvious scatter that you get when pooling the observations of a group of people. Such light curves clearly show that the 0.1 magnitude accuracy thing is a kind of idealized limit, which can more likely be approached, but not very often actually achieved. I thus arrived a little late at considering the possibility of becoming a variable star observer, and it turned out to be quite an adventure.

I managed to get hold of one or two amateur books on variable stars and variable star observing, and the very first thing that struck me about this area of amateur astronomy was that it isn't just "amateur astronomy," it is amateur astrophysics. The observations made by amateur variable observers are real data in the truest scientific meaning of the word, and to be honest, I found it astounding that such simple observations could reveal things going on inside distant suns that are so far away that they can only be seen as points of light. When you have this kind of "revelation," the often spoken of "addictive" quality of variable star observing comes as no surprise, but in addition to being able to make scientifically valuable observations on a regular basis, I know that in my own case I wanted to know more about what really does go on within stars, to make them vary as they do – or indeed, not vary at all.

After leaving high school I got a degree in physics at my local university, and while this helped in my wish to know more about the physics of stars, the fact is that much of stellar astrophysics is a specialty unto itself and not the kind of stuff that you are likely to come across in a straight physics degree course. However, as a result of my new found interest in variable stars, I got to know a much more experienced amateur variable star observer who did photoelectric photometry with a real live photomultiplier ("live" being very appropriate here, because the high-voltage power supply that ran his photomultiplier was housed in a washing-up bowl, which sat on his rather damp lawn). This guy also did some of his own data analysis, and he was certainly the kind of person that any novice variable star observer was truly privileged to know. I remember him, though, complaining on more than one occasion about the lack of decent books on both variable star physics and for that matter on just stellar physics itself, which were suitable for amateurs. He himself had to pick up what effectively were disjointed fragments of information from professional research papers, specialist monographs, and the occasional textbook. He just happened to be a science librarian, which was very fortunate at a time when there was no Internet. Even these days, much of this kind of information still very often comes in the form of articles – online or otherwise, which just don't have the space to be able to deal with a subject in the kind of depth that it maybe deserves, or even worse, it gets the odd paragraph or two in either a more general astronomy book or in books that are specifically written as practical observing guides. There are still also, of course, the student textbooks and the research papers.

It goes without saying that textbooks on astronomy are not written with amateur astronomers in mind. The fact is that any student who wishes to become a professional observational astronomer has to learn a lot of background theory – "required reading," as they say, and there has always been a plentiful supply of textbooks, some of them veritable classic works to give students what they need. Where does a serious amateur astronomer get his or her background theory from, though? I'm sure that many amateur astronomers probably have a sufficient background in mathematics and physics to be able to tackle at least the more basic level textbooks – but then again, there will be many that don't. There will surely also be many who would say that for the work they do, they simply have no need for this kind of theoretical background – but wouldn't it be nice to have it anyway, especially if it could be made more accessible and didn't require a higher education level background in physics and mathematics?

This book is written for those amateur astronomers who would be inclined to answer "yes" to this question and who do not have said background in math and physics. Here in the early 21st century, amateur astronomers are uniquely placed in terms of technology in the form of CCD cameras and computers. Also, there are many resources – particularly in the form of the Internet – to be able to carry out truly rewarding and fulfilling programs of astronomical research, which the whole global astronomical community, both amateur and professional alike, will want to know about. In this grand scheme of things, a basic knowledge of stellar astrophysics will surely find its place. It has to be said, though, that what we present here are really just the basics, but which nonetheless deal with many of the kinds of topics that would be required reading for an undergraduate astronomer. The difference is that here we avoid the kind of mathematical rigor required of the student, while at the same time hopefully ensuring that the physics and the astrophysics remain clear and concise.

Stellar astrophysics is an enormous subject with many specialist areas for which, unfortunately, there just isn't space here to go into in the depth that they might deserve. Such areas include binary stars and indeed variable stars themselves. One exception, though, is our discussion of stellar pulsation, which is itself a wonderful illustration of the kinds of things that go on inside stars. The discussion of many of the topics included in a book on this subject inevitably involves describing some aspect of spectroscopy, and we have done this here when the need has arisen. Spectroscopy, though, is so important to all areas of astronomy, including stellar astrophysics, that it certainly does merit a separate book, which can give it a more in depth treatment. The present book then can certainly be regarded as a companion to the author's *Spectroscopy – the Key to the Stars*, also published by Springer.

Stellar physics is basically all about learning to interpret and understand the information that is contained in starlight. In many ways starlight is like a river. As astronomers, we sit facing the mouth of that river. Just as with a river here on Earth, where a sample of water can reveal to the Earth scientist a great deal about the river's journey from the mountains to the sea, so, too, can the starlight that enters the objective end of your telescope tell the story of its long journey from its source in the heart of a distant star, through the star's outer layers, across interstellar space, and down through Earth's atmosphere.

As already indicated, behind the astrophysics there lies a fair bit of pure physics, which, whenever it is needed, we will be sure to introduce from scratch, so that even if your memory of high school physics lessons is growing somewhat dim, there should be no problem. Perhaps the boldest step we've made is to introduce some very basic stuff about numbers right at the start (all right, extremely basic mathematics if you insist); it really is harmless stuff, which should not cause any distress. You will in fact find that being able to input a number into a simple equation in order to produce another number, which is able to tell us something significant about stars, is very satisfying, and of course we'll give detailed step by step instructions each time on how to do this with a pocket calculator.

The result of being able to make use of some very simple mathematical equations together with a little knowledge of some basic physics will, as we shall see, take us a very long way in our understanding of the astrophysics of stars. Much of what we observe and know of stars will then seem to be the natural and logical result of basic physical processes going on within them, and also in the space that lies between them and us. Finally, you'll also become familiar with the meaning of many of the ideas and terms used a lot by stellar astronomers. These include things such as color indices, color excess, optical depth, absorption, scattering, and many more, which, if nothing else, might make the business of going through "that research article" just a bit easier. Let's begin our journey, then, along the river of starlight, by becoming familiar with a few numbers.

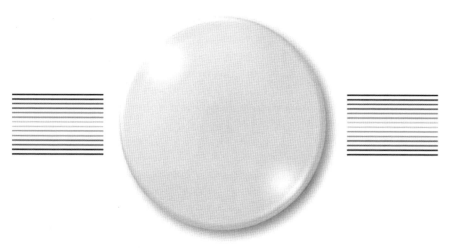

Starlight by Numbers

They say that mathematicians drink a toast which goes: "Here's to pure mathematics – may it never be of any use to anyone." Well by that score, I'm definitely not a mathematician; at least not a pure mathematician. Let's face it, for many people (perhaps myself among them) mathematics reaches the parts of the brain that hurt, so when we do seek to solve a mathematical equation as we will from time to time, you can be sure that there's a real reason for doing this.

One obvious reason is that the number that results from solving an equation may be of real importance to us; a less obvious reason, but one that is just as important and perhaps even more important to the learner is that a simple equation can be used to explore some part of astrophysics. The basic idea is to use a pocket calculator to try out or to plug different numbers into the equation; this enables you to get a "feel" for the kind of numbers that are involved in the solution to the equation (are they huge numbers or very small ones for example?). This process of "equation exploration" will also show you how the all-important solution to the equation actually depends on the different numbers that get plugged in. For example, will doubling an input number simply double the value of the answer or maybe multiply it by four. The result is that by doing this kind of thing you are guaranteed to gain a much deeper understanding of that particular bit of astrophysics.

You can if you wish ignore the equations we encounter without really losing anything, but if you have a calculator, then do have a go at using it to explore an equation; you'll soon come to realize just how valuable and even enjoyable this is. As for the kinds of equations that we will come

K. Robinson, *Starlight*, Patrick Moore's Practical Astronomy Series,
DOI 10.1007/978-1-4419-0708-0_2, © Springer Science+Business Media, LLC 2009

across, have a look further down at Equation (10); if an equation like this presents you with no problems, then feel free to go to the final section of this chapter on "Star Distances by Numbers." The main purpose of this fairly short chapter is to show you how to solve equations such as this and thus hopefully give you a solid foundation and a smooth read through the rest of the book. For any other mathematical points that come up, we'll deal with them only when the need arises in order to prevent you from getting "mathematical indigestion." So here goes, starting with some very basic stuff about numbers.

Large Numbers and Small Numbers

Start with the number 100; a "1" followed by two zeros, which of course also equals 10×10, or two number 10s multiplied together. Similarly the number 1,000 – a "1" followed by three zeros is the same as three number 10 s multiplied together. The way that mathematicians and scientists write a number like, for example, 100,000 is 10^5. This is a shorthand way of writing the number "1" followed by five (5) zeros or five number 10 s multiplied together; so "100" becomes 10^2 and "1,000" becomes 10^3. This clearly avoids the need to write long strings of zeros, but it does more, as you might expect. One way of saying the number 10^3 (besides saying "one thousand") is of course "ten cubed," but a more precise way is to say "*ten to the power three*" or just "ten to the three," and then, for example, 10^5 can be spoken of as "ten to the power five" or "ten to the five," and so on. The process of taking a quantity "*x*" of the same number and multiplying them together is called *raising the number to the power "x"* and in particular, numbers such as 10^7, 10^8, etc., are often referred to as *powers of 10*. The actual number "*x*" – for example, the "5" in 10^5 – is called the *index of the power*, or just the index and the plural of index here is *indices*.

The Rule of Indices

If we multiply 100 by 1,000, we get 100,000, or using our new "*powers of 10*" notation,

$$10^2 \times 10^3 = 10^5 \tag{1}$$

So when we multiply two different powers of 10 together, we simply *add the indices together* to get the resulting power of 10. This is a very important and powerful rule in mathematics called the *rule of indices*, and it can be applied to numbers other than 10. For example, a very important number that is used a lot by both astronomers and physicists is the number 2.718 (to 3 places of decimals). Mathematicians give this number the symbol "e" just like they give the number 3.142 the symbol "π." So, for example,

$$e^7 \times e^5 = e^{12} = 2.718^7 \times 2.718^5 = 2.718^{12} \tag{2}$$

Provided the number whose powers are being taken (in this case "e" or 2.718) is the same throughout the equation, the rule works. The number 2.718^{12} is very large, by the way, and we'll see shortly how to write such a number, but first let's extend this powerful rule of indices.

What about the number 10 itself? It's simply the number 1 followed by one zero, so we should be able to write it as 10^1. We can check that this is okay by making sure it satisfies the rule of indices; so, for example, 10 $\times$ 100, which equals 1,000, can also be written $10^1 \times 10^2 = 10^3$; and yes, the indices do add together correctly. Also by virtue of our example using the number 2.718, we can say that any *number raised to the power "1" is just the number itself*; so "e^1" just equals "e." What about the number "1," or 1 followed by no zeros? In the powers of 10 notation this would be written 10^0, and this too satisfies the rule of indices because, for example, $10^0 \times 10^2 = 10^2$, which is the same as saying 1 $\times$ 100 equals 100. Once again the rule extends to all numbers so that *any number raised to the power zero is equal to* 1; so again, for example, $e^0 = 1$.

With what we've learned so far we can make very large numbers by raising a smaller number such as 10 or "e" to a very high power. But what about very small numbers? Start with the number 100,000 or 10^5; If we divide this by 100 or 10^2, we get 1,000 or 10^3. In other words,

$$\frac{10^5}{10^2} = 10^3 \tag{3}$$

So when we *divide* one power of 10 by another we have to subtract the index at the bottom; i.e., in the denominator from that at the top in the numerator. Another way to write Equation (3) is like this:

$$\frac{10^5}{10^2} = 10^5 \times \frac{1}{10^2} = 10^3 \tag{4}$$

So here we've turned the division of two powers of 10 into the multiplication of one power of 10 with another number that involves the reciprocal (the reciprocal of any number simply equals the number 1 divided by that number) of a power of 10. This has to satisfy the rule of indices and the only way that it can do this is to make $1/10^2$ equal to 10^{-2} because then we get

$$10^5 \times 10^{-2} = 10^3 \tag{5}$$

The indices check out because $5 + (-2)$ is the same as $5 - 2$, which equals 3. This has also told us that a small number such as $1/100,000$, or $1/10^5$, is written as 10^{-5}. Extending the idea again to our friend the number 2.718 or "e," the reciprocal of 2.718 or $1/e$ would be written as e^{-1}; it equals 0.368.

The Rule of Indices for All Indices

Any number can in fact be raised to a power that does not have to be either a positive or a negative whole number; an important example of this kind of power would be the number $x^{1/2}$. We can easily see the meaning of this number by multiplying it by itself and applying the rule of indices because then we get; $x^{1/2} \times x^{1/2} = x^1$, which just equals x. So $x^{1/2}$ is just the square root of x, and using the same procedure, $x^{1/3}$ is the cube root of x and so on.

A trickier problem is the meaning of something like $x^{3/8}$. We can in fact "kill two birds with one stone" here by thinking about a number such as $(10^5)^3$. Notice that this is not the same as $10^5 \times 10^3$, which would of course equal 10^8. Instead, this is the number 10^5 multiplied by itself 3 times – in other words, it's the number 10^5 *raised to the power* 3 (a number raised to a power, which is then itself raised to some other power). The number 10^5 multiplied by itself 3 times is the same as $10^5 \times 10^5 \times 10^5$, which of course equals 10^{15}. See how the number 15 is just equal to 5×3? So if we have a number that is raised to some power and *we raise it again* to some other power, we multiply the two powers together to get the final answer. So in general terms; $(x^y)^z$ is equal to $x^{y \times z}$ or just x^{yz}. This idea in fact extends to any number of indices, so for example $((e^2)^3)^4$ is equal to e^{24}. We see now that $x^{3/8}$ is the same as $(x^{1/8})^3$; i.e., the eighth root of x multiplied by itself 3 times. We shall use this important application of the rule of indices in chapter *A Star Story – 10 Billion Years in the Making*, where we need to be okay with the fact that $(x^3)^{1/2}$ is the same as $x^{3/2}$ or $x^{1.5}$.

Finally we can have numbers like $e^{-0.43}$; i.e., $2.718^{-0.43}$. This, however, is not the kind of thing to try and visualize in any way, nor to try and work out with pencil and paper. We shall find a need to be able to work out this sort of thing in chapter *Space – The Great Radiation Field*, and the best way is to find the key on your calculator labeled "x^y" or maybe "y^x." (If your calculator is not a scientific one then do give serious consideration to purchasing one – it will become your great friend.)

Try, for example, tapping in the number 2.512, then press the "x^y" key; now tap in the number 2.4 and finally press the "=" key to get the answer 9.121. You've just calculated the ratio of brightness for two stars whose magnitudes differ by 2.4.

Fortunately, working just with powers of 10 is much simpler, but the need to do so crops up all the time, so it pays to be comfortable when using them. Following are a few examples to illustrate how things work.

Working with Powers of Ten

So far we've learned how to multiply together two powers of 10; so for example

$$10^8 \times 10^5 = 10^{13} \tag{6}$$

This kind of operation can be extended to any number of terms on the left-hand side, so, for example

$$10^8 \times 10^5 \times 10^3 \times 10^7 = 10^{23} \tag{7}$$

A quantity such as $10^{11}/10^4$ can also be written $10^{11} \times 10^{-4}$, which of course equals 10^7, but note also that you may come across numbers such as $10^{-11} \times 10^4$, which in this case equals 10^{-7} ($-11 + 4 = -7$). Finally, consider an expression such as $\frac{10^{11}}{10^{-4}}$ This is equivalent to $10^{11} \times \frac{1}{10^{-4}}$. Go back to old habits and think of the number 10^{-4} as one ten thousandth, and ask "How many times does one ten thousandth go into 1?" The answer is, of course, ten thousand times, or 10^4. So, a number such as $1/10^{-4}$ becomes 10^4. The general rule here is that an index in the bottom line or denominator of an equation can be simply moved up to the numerator or top line *provided you change the sign of the index*. So the above expression becomes

$$\frac{10^{11}}{10^{-4}} = 10^{11} \times 10^4 = 10^{15} \tag{8}$$

We can now put these ideas together; for example, $\frac{10^{-34}}{10^{-7}} \times 10^8$ becomes $10^{-34} \times 10^8 \times 10^7$, which equals 10^{-19}. The key to all of this is making sure that the signs (+ or −) of your indices are correct and that you add all the indices *algebraically* – i.e., you take the sign of each index into account.

Scientific Notation

Of course, most numbers do not consist simply of multiples or powers of 10; here, for example, is a fairly large number, 299792.0. This is, in fact, the speed of light in kilometers per second, and note that we've included the decimal point. If we divide this number by 10, the decimal point moves one place to the left to give us 29979.2, and dividing again by 10 we get 2997.92, giving us the speed of light divided by 100.

To get our speed of light back we could simply reverse the process and move the decimal point two places to the right, but another way to write the restored number would be 2997.92 × 100, or 2997.92×10^2. Pursuing this idea further, we can write the speed of light as 2.99792×10^5 km/s. This is the standard way to write down the speed of light (or indeed any large number) in what is called *scientific notation*. A very small number such as 0.0000005 can be written as 5.0×10^{-7} in this notation by multiplying by 10^7, thereby moving the decimal point 7 places to the right and then multiplying by 10^{-7} to give us back the original number. This number is approximately equal to the wavelength of green light in meters (more on this in chapter *From Light to Starlight*). The number 2.718^{12}, which we encountered earlier, can be evaluated using your calculator by entering 2.718 followed by pressing the "x^y" key followed by the number 12, and finally followed by the "=" key to give the number 162552.416. In scientific notation this number is written as 1.62552416×10^5.

Far from being just an esoteric way to write down large and small numbers, scientific notation, as we'll now see, makes it much easier to get a useful number out of an equation if the numbers we plug in are in this form. Here in normal number form is an equation that tells us how much energy is carried by the green light we just mentioned:

$$E(energy) = \frac{0.0000000000000000000000000000000006626 \times 299700000}{0.0000005}$$

(9)

This is not a pretty sight; but when the numbers are written in scientific notation it looks like this:

$$E = \frac{6.626 \times 10^{-34}}{5.0 \times 10^{-7}} \times 2.997 \times 10^8 \tag{10}$$

To get the actual value for the energy out of this equation is very straightforward; we divide the whole thing into two parts; one consisting of the ordinary numbers like the 6.626 and the other part consisting of the powers of 10 terms. Now "E" becomes

$$E = \frac{6.626 \times 2.997}{5.0} \times \frac{10^{-34}}{10^{-7}} \times 10^8 \qquad (11)$$

We can now deal with this one a bit at a time. First use your pocket calculator to solve the "normal numbers" part and get the answer 3.972 to three decimal places, then put this number "to one side" while we do the powers of ten part. Using the rule of indices, we can put the 10^{-7} onto the top row by changing the -7 to $+7$ and then we just have to carefully add the indices together, taking into account their respective signs; this gives us

$$10^{-34} \times 10^8 \times 10^7 = 10^{-19} \qquad (12)$$

All we have to do now is to multiply this by the first bit of the calculation (the 3.972) to get the final answer – 3.972×10^{-19}.

It can often happen when doing a calculation this way that you end up with an answer that (just as an example) may look something like this: 122.434×10^7. You can if you wish leave it like this, but if you're a fussy sort of person then you should rearrange it to look like this: 1.22434×10^9.

If you're unfamiliar with doing calculations in scientific notation then using the above method exactly as described is good for practice; it gets you familiar with handling powers of 10 and with the way in which scientific notation works. However you can, if you wish, enter a number in scientific notation directly into your calculator. To do this with, for example, the number 6.672×10^{-11}, first enter 6.672 in the normal way. Then press the key marked "Exp" on your calculator; you'll now have two small zeros at the top right corner of your calculator display. Now enter the number 11; this will appear instead of the two zeros. Finally, press the key marked "+/–;" this will turn the "11" into a "–11." If you now want to multiply or divide this number by another scientific notation number simply press the "×" or the "÷" and enter the new number, remembering, of course, that if this number involves a positive power of 10 then there's no need to use the "+/–" key. Happy calculating!

So now we've calculated the amount of energy carried by light having a wavelength of 5.0×10^{-7} m, which is equal to 3.972×10^{-19}. But 3.972×10^{-19} what? The quantity of energy that we have calculated here is measured in units called *joules*, and judging by the fact that the answer we got is a very tiny number, maybe we should ask ourselves if the joule is

perhaps too large a unit of energy for the kind of situation we are dealing with. On the other hand it may be that green light simply doesn't
carry much energy and leave it at that. We'll say more about this kind of
thing as the need arises; the important thing here is that you can handle
a calculation like Equation (10) and be confident of getting the correct
answer.

Star Distances by Numbers

One of the most important things that can be known about any star is its distance. This precious number enables astronomers to determine the true brightness of a star and thus its total power output, or *luminosity*. It can even enable an estimate to be made of the star's temperature. We'll see in due course how these things are done, but for the moment and really for the sake of completeness, we'll say a word or two about the meaning of the numbers, which are used when discussing the distances to the stars.

As amateur astronomers we fairly quickly get to know that, when talking about distances out beyond the Solar System we use the *light year* – the distance that a beam of light travels in 1 year. Light travels at a speed of 2.998×10^8 m/s, and the length of a "year" is 365.242 days (to three decimal places), or 3.15569088×10^7 s, and so it is a fairly straightforward calculation to work out that a light year is the mighty distance of about 9.461×10^{15} m. Even so, the distances to the visible stars are of the order of around 10 to around 1,000 light years, with many stars in our galaxy being at much greater distances than this, of course. Professional astronomers, though, tend to use an alternative star distance measure – the *parsec*, or "pc" for short, which stands for "parallax second."

The parsec comes straight out of the only direct way to measure the distance to a star, which involves determining with the utmost care the apparent change in position of the star against a background of more distant stars (a background of distant galaxies is even better) over a period of six months – i.e., as Earth swings between opposite sides of its orbit around the Sun, as shown in Fig. 1. Again as amateur astronomers we learn very soon that the only way to measure the "distance," or separation between any two points in the sky, is in terms of the angle between two lines drawn from these points, which intersect at our eyes as shown in Fig. 2.

Using this method, the angular separation between the two pointer stars of the Big Dipper (the Plough in the U.K.) is about 5°; the angular sizes of the Sun and Moon are both about half a degree, or 30 *arc minutes*. The apparent diameters of the planets are of the order of a few tens of *arc seconds* (or "arcsec," for short), where 1 arcsec is 1/3,600th of a degree. By the time we get to the separations of close double stars, we're talking of the order of maybe a few arc seconds, and it's well known to amateur observers that one of the most demanding tests for a telescope is its ability to separate or resolve close double stars.

The half yearly shift of even relatively nearby stars amounts to less than 1 arcsec, and so determining these shifts puts even large Earth-based telescopes at the very limits of their performance. The result is that until the *Hipparcos* satellite considerably improved things, it was very much the norm for star distances to be accurate only to about 10–20% and the limiting distance was about 300 light years, or about 100 pc.

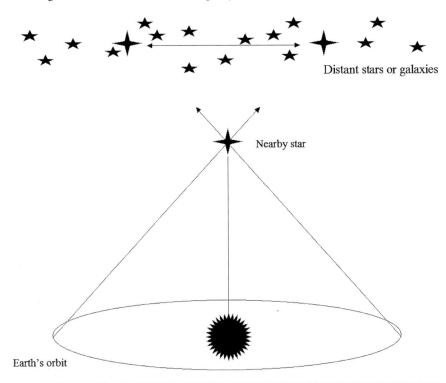

Figure 1. The apparent position of a nearby star shifts against a background of more distant stars or galaxies over a period six months as Earth swings between opposite points in its orbit. By measuring this (tiny) shift and knowing the Earth–Sun distance, simple trigonometry enables the distance to the star to be determined. In reality, the process is rather more involved than this simple diagram would suggest.

The actual *parallax* of a star is defined to be the angle between two lines running from the center of the Sun and a point in Earth's orbit whose distance from the Sun is 1 astronomical unit (A.U.) (1 A.U. is the average Earth–Sun distance, which is equal to 1.496×10^{11} m) and which intersect at the star itself (as shown in Fig. 3). Thus the parallax is equal to half of and not the whole apparent angular shift of a star over a period of six months. So the parallax is the angle at the apex of an extremely thin

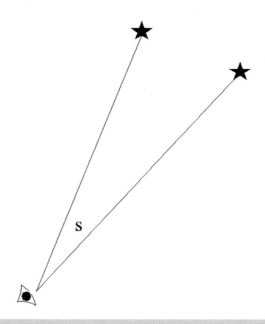

Figure 2. The apparent "separation" of two objects on the sky (this could also be the apparent shift of one star over a period of six months) is measured by the angle they subtend at the eye (or telescope). This angle is measured in degrees, minutes (or *arcminutes*), and seconds (or *arcseconds*).

right-angled triangle. The length of the base of this triangle is just 1 A.U., and if the parallax of a star were in fact equal to 1 arcsec (1/3,600th or 2.778×10^{-4} degrees), then its distance, using high school trigonometry, would be equal to 1 A.U. divided by the tangent of 1 arcsec. If you try this on your calculator you should get the answer 3.086×10^{16} m. *This distance is defined to be equal to 1 parsec,* and if we divide it by the number of meters in 1 light year, then we can see that 1 parsec is equal to 3.262 light years.

The advantage of using parsecs is that if you know the parallax of a star, then its distance in parsecs is just equal to the reciprocal of the parallax. So, as we've seen, a parallax of 1 arcsec corresponds to a distance of 1 parsec; a parallax of 0.5 arcsec results in a distance of 2 parsecs and so on. So, for example, the parallax of Proxima Centauri the nearest star is about 0.75 arcsec; the reciprocal of this is 1.333, which is the distance to Proxima in parsecs that, when multiplied by 3.262, gives us 4.35, or its distance in light years. The reason that this simple relationship works is because the baseline of these "parallax right-angled triangles" all have the

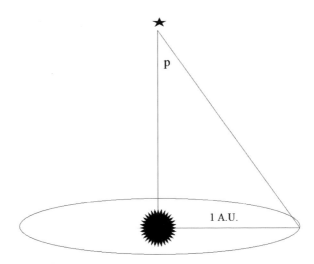

Figure 3. The angle "p" defines the "parallax" of a star; it is equal to half (not the whole) of the annual apparent shift of the star's position. This tiny angle thus forms the apex of a very narrow right-angled triangle whose base has a length of 1 Astronomical Unit.

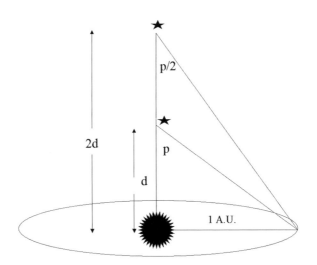

Figure 4. The use of right-angled triangles makes parallax trigonometry very simple. If the distance to a star is doubled, its parallax is halved; triple the distance, and its parallax becomes one third, and so on.

same base length, i.e., 1 A.U. (they are, in fact, what mathematicians call similar triangles), and so if the distance doubles, the parallax halves, etc., as shown in Fig. 4.

So there we have it; this is pretty well all the mathematics you'll need for now (oh! and do remember that number "e" or 2.718, which we'll meet again later) to hopefully get that extra bit out of this book. Read on!

Key Points

- When the same number is multiplied by itself "x" times, the number is said to be raised to the power "x."

- The number "x" is called the index of the power, or simply the index (plural "indices").

- If a number that is raised to the power "x" is multiplied by the same number raised to the power "y," then the result is the same number raised to the power "$x + y$;" the indices simply add together, and this is called the rule of indices.

- For any number "z," "z^1" equals "z" and "z^0" equals 1.

- For any number "z" and index "x," the number "z^{-x}" is equal to the reciprocal of z^x; i.e., $1/z^x$.

- Any number such as $(x^y)^z$ is equal to x^{yz}, i.e., when a number is successively raised to several different powers, the final result is equal to the number raised to the *product* of all the powers.

- Very large and very small numbers are usually written in scientific notation. For example, 5,000 is written as 5.0×10^3 and 0.005 is written as 5.0×10^{-3}.

- The number "e," which equals 2.718, is very important and should be remembered.

- A star at a distance of 1 parsec, or 1 pc, would have a parallax equal to 1 arcsec; no star is as close as this.

- The distance in parsecs to any star whose parallax is known is simply the reciprocal of the parallax.

From Light to Starlight

During my final year at high school our physics teacher, Mr. Taylor (Mr. Taylor also ran the truly excellent school astronomical society) told us that, despite the fact that by that stage we'd learned a lot about electromagnetic waves and photons, at the end of the day physics could not actually say exactly what light is. This is, of course, true, but remarkably it doesn't matter (well, it might matter to a philosopher), provided we accept two very important things about physics. First, what we generally call the great theories of physics, such as classical electromagnetic theory and quantum mechanics, are in fact models, which physicists use to try and explain as best they can the results of experiments and hopefully to predict the results of future experiments. Second, while it is often the case that one theory or model (in this case quantum theory and photons) supersedes an earlier one (the idea of electromagnetic waves), the older theory is very often still extremely useful.

For example, few people these days would subscribe to the idea that Earth is truly fixed at the center of the universe as prescribed by most ancient astronomers. However, as modern day amateur and professional astronomers, we still make frequent use of the idea of the celestial sphere upon which stars and planets, etc., appear to move around a fixed central Earth; so here's an example of an "old dog" that *doesn't* necessarily need to learn new tricks.

Astronomers, too, use these theories of physics to try to explain why, for example, the light from a star has the properties that we observe. In this chapter, we'll look at some of the most basic properties of light using

K. Robinson, *Starlight*, Patrick Moore's Practical Astronomy Series,
DOI 10.1007/978-1-4419-0708-0_3, © Springer Science+Business Media, LLC 2009

the great theories that have been developed to explain them. What's more, as indicated above, we'll be quite happy to make use of the older idea – namely, that of electromagnetic waves, when it seems that this gives a perfectly reasonable and maybe even an easier to understand explanation of things. Then, when necessary, we'll use the more modern idea – that of quantum theory and photons – when the old theory just won't do. This is what the professionals do; so if they can, we can!

Let There Be (an Idea About) Light

Despite the fact that Isaac Newton favored the idea that light consisted of a stream of energetic corpuscles, the first great theory about light dealt with waves. The whole thing was, in fact, a magnificent coming together of experiment and theory. Experiments showed that light was clearly some form of wave motion, but then, like setting the capstone on top of one of the great pyramids, James Clerk Maxwell showed theoretically that electric and magnetic fields could travel together as waves through space. What's more these *electromagnetic waves* traveled at a speed that was equal to the then-known speed of light itself. *Voila!* Light was an electromagnetic wave.

Having made such a bold statement, let's see what we can do with this tremendous idea. The thing that probably causes most difficulty with understanding waves is that they are dynamic entities. They are by their very nature changing constantly in time. Also where light waves are concerned, we have to deal with the rather abstract idea of electric and magnetic fields. In addition, the mathematics of wave motion doesn't come cheap, but at least we don't have to go down that particular road. We will in due course, though, need to get our heads round some numbers, but to start with let's establish a few basic facts about waves.

The first thing to do is to take out the time element by imagining that we can "freeze frame" a wave so that we can poke around and examine it. The second thing is to take the kind of wave that we are familiar with and the obvious example here is a water wave. Drop a stone into a pond, and the water wave spreads out as a series of circular ripples. Now "freeze" this water wave in time; we see that the "frozen" wave consists of circular areas where the water level is relatively high, alternating with areas where the level is relatively low. The high zones are called *crests* and the low zones *troughs*. Most importantly, if we measure the distance between two neighboring crests, we find that it's the same wherever we are on the wave. This all important number is called the *wavelength*, and it is always represented in the literature by the Greek letter lambda or "λ." For water waves, the wavelength may be measured in centimeters for a pond or from meters to kilometers for an ocean.

An even simpler example of a familiar wave is one we can produce using a length of rope; we tie one end of the rope to a post and then wiggle the free end up and down. Again, if we make a movie of this and look at an individual frame, we would see that the distance between two neighboring crests, or the wavelength of our rope wave, was the same all the way along the rope. If we want, we can vary the speed at which we wiggle the free end of the rope. If we do this, we would see that the more rapidly we wiggle the free end of the rope the shorter the wavelength becomes, whereas a more leisurely wiggling action produces a longer wavelength. We can, in fact, be more precise about our wiggling action; by a rapid wiggling action we mean that we move the free end of the rope up and down more rapidly, and we can also define "one wiggle" as one complete up and down movement of the end of the rope. Clearly, a more rapid wiggling action will result in more "wiggles" in a given interval of time – say, 1 s. In fact, the total number of complete wiggles that take place in 1 s is called the *frequency* of our rope wave, and indeed of any kind of wave we care to think about – it's always denoted by the Greek letter nu or "ν" in the literature. So more wiggles per second means a higher frequency, but as we've seen rapid wiggles make for a shorter wavelength.

If we were to watch our rope wave movie, we'd see that the wave shape appears to move along the length of the rope, and if we had tied a piece of rag to the rope at some point along its length, we'd also see that as the wave shape moved along, the piece of rag would move up and down as crests and troughs in the wave came by. This motion of a point on the wave at right angles to the direction along which the wave pattern moves means that our rope wave is an example of what's called a *transverse wave*. The maximum distance traveled by the piece of rag away from the mean level of the rope is another important quantity associated with waves called the *amplitude,* which measures the height of the crests or the depth of the troughs. Figure 1 shows the basic "anatomy" of a transverse wave.

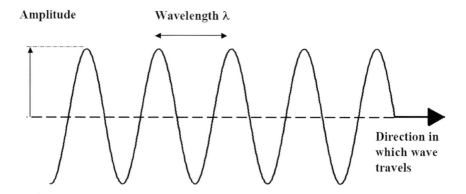

Figure 1. The basic "anatomy" of a transverse wave defines the wavelength as the distance between two neighboring crests and the amplitude as the maximum height of a crest above the mean level of the wave.

Another thing we'd see by carefully watching the movie is that one complete up and down movement of the piece of rag corresponds exactly to one single wiggle of the free end of the rope. If the rag starts at the top of a crest in the wave, it will descend, as the wave moves along, to the bottom of a trough and then back up again as the next crest comes along. So after one wiggle, the wave pattern in the rope has moved along by one wavelength, i.e., by a distance of λ as shown in Fig. 2. After ν wiggles the wave will move along by a distance of $\nu \times \lambda$; ν is the number of wiggles in 1 s. So in 1 s the wave moves a distance of $\nu \times \lambda$; the distance the wave moves in 1 s is of course the wave's velocity V and so

$$V = \nu \times \lambda \qquad (1)$$

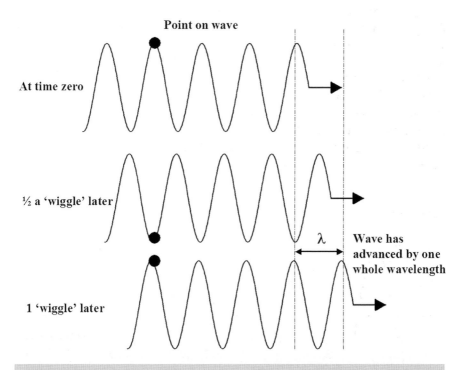

Figure 2. After one "wiggle," i.e., one complete up and down motion of any point on the wave, the wave pattern advances by one whole wavelength "λ"; after "ν" "wiggles," i.e., after 1 s, the wave advances by a distance $\nu \times \lambda$, and so this distance that is covered in 1 s equals the wave's velocity "V."

This simple but exceedingly important formula applies to any wave, including light waves. For a given type of wave, the value of the velocity V is fixed, so this formula confirms what we said above, that if, for example, we increase the frequency, the wavelength must decrease, i.e., become shorter in order to keep the velocity constant.

As you might expect a light wave is more subtle than either a water wave or a rope wave. In these latter cases, it's obvious that something is moving up and down as the wave pattern moves along; not surprisingly, then, once experiments showed that light behaves like some form of wave, it was of great concern to physicists at the time as to what if anything might be moving, or, in other words, what would be meant by the "amplitude of a light wave?"

In fact, it turns out that the only thing that "moves" in a light wave is the wave pattern itself, and this as you'd expect moves along at the speed of light (always denoted in the literature by a small letter "c"). Clearly, though, there must be something that we can measure, which is changing as the light wave moves along, and once more let's imagine that we can freeze frame a light wave so that we can examine it. This time, however, we can't "see" the light wave, but if we know where the source of the light is and by means of a detector (a CCD camera, for example) we know where the light wave ended up (and trusting that light travels in a straight line), we can at least see the line along which the light wave is traveling.

What we do now is carefully move along this line with a compass. What we'd see is remarkable; at first, the compass needle would point in one direction, and this direction would be at right angles to the light wave's line of travel. Then, as we move along, the needle would flip around and point in the opposite direction. Further along still it would flip around yet again, and eventually we'd realize that these compass needle flips happened at equally spaced intervals along the light wave's path. This tells us that a magnetic field exists along a light wave, and what's more, this field periodically changes direction along the length of the wave.

Hardly able to contain our excitement, we invest in a magnetometer – a device that can tell us how strong a magnetic field is as well as its direction. We carry out the same procedure again, this time measuring the strength as well as the direction of the magnetic field as we move along the "frozen" light wave. Armed eventually with a mass of data, we can now plot on a graph the strength and direction of the magnetic field against distance along the light wave; our graph would itself have a beautiful wave shape.

So, unlike water waves and rope waves, light waves don't involve any sideways movement of something in space. Instead, it is the strength and direction of a magnetic field that varies as the wave moves along. At any given point on the wave, the magnetic field strength increases from zero to a maximum value, decreases to zero again, and then increases again but pointing in the opposite direction.

There's something else, too, that comes straight out of Maxwell's classical electromagnetic theory. A magnetic field that varies as it does here does so because of an electric field that is varying in a similar way; so our light wave also involves a varying electric field. The direction of the electric field is at right angles to both the magnetic field and to the direction along which the wave travels as illustrated (rather crudely) in Fig. 3, and so we have an *electromagnetic wave*, which, as with water and rope waves, is an example of a transverse wave. The wavelength of an electromagnetic

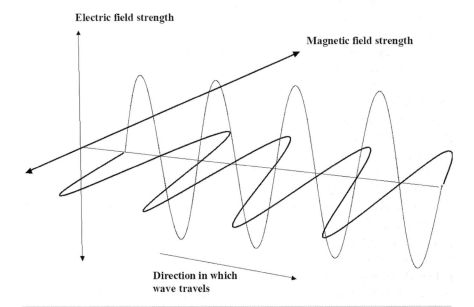

Figure 3. This is, it must be admitted, a pretty simplistic representation of an electromagnetic wave, i.e., a light wave. It does at least show that the directions of the electric and magnetic fields are both at right angles to each other and to the direction of the wave's travel. It also shows that the electric and magnetic waves are "in phase," i.e., crest matches crest and trough matches trough all the way along the wave. Clearly, trying to represent a light wave that spreads outward in all directions like this would result in a very messy diagram.

wave is the distance between two neighboring points where the electric or magnetic field strengths are a maximum (the distance is the same in both cases), and the amplitude of the wave is now measured not by a physical distance but by the maximum strengths of the electric and magnetic fields themselves.

At this point, it has to be admitted and most of you have probably realized that for visible light (even in this era of multi-wavelength astronomy we still tend to think of visible light first), the wavelengths involved are very tiny indeed, and no way could we measure them with a simple compass. The idea behind this thought experiment, though, is completely sound and fits in perfectly with the classical wave theory of light. In this theory, the amplitude of the light wave, i.e., the maximum strengths of the electric and magnetic fields, determine how *bright* the light is and the wavelength (and by implication from Equation (1), the frequency) determines the *color* of the light.

Electromagnetic waves (including visible light) can have a whole range of wavelengths; a range of wavelengths constitutes a *spectrum* – in this case, the *electromagnetic spectrum,* or the "e-m" spectrum. We can sometimes see the visible part of the electromagnetic spectrum spread out in the sky for us in the form of a rainbow, or, if we can't be bothered to wait for it to rain, passing sunlight through a glass prism will do the same job.

A very obvious but nonetheless very important feature of both rainbows and the spectrum of sunlight produced by a prism is that there are no noticeable gaps between the colors. For example, there's no gap between the orange and the yellow or between the blue and the green. This suggests to us that sunlight contains not just some but all of the colors or wavelengths that make up the visible part of the e-m spectrum. Obvious though this is, it is important, because when the spectra of the Sun and most stars are studied in greater detail, it is seen that there are, in fact, "colors" that are missing. The shortest wavelengths in the visible spectrum are perceived by us as the color violet; the longest wavelengths are perceived as deep red, and all the other colors that we can see have wavelengths somewhere in between. There are also, of course, other "colors" that we can't see.

Nature's Color Palette – The Electromagnetic Spectrum

An artist will take basic colors from the visible spectrum and mix them together to give all possible shades, tints, and hues that we see around us in nature. Alas, what even the greatest artist cannot do is to take nature's extra hidden "colors" that come from electromagnetic waves of other wavelengths and add them to his or her landscape painting.

By contrast, one of the great wonders of modern astronomy is that it has been able to do just this – to give us a truly awesome view of the great river of starlight that enters our telescopes. Most amateur astronomers observe the visible part of the spectrum, but as we shall see later, nature's extra "colors" play a crucial and fascinating role in determining what we see in visible light. Understanding the role played by these invisible parts of the e-m spectrum enables the amateur astronomer to gain a much deeper appreciation of his or her observations.

The wavelength λ determines the kind of electromagnetic radiation we are dealing with; it's a number that can be directly measured, as we'll see shortly, but first let's look at the kind of numbers that are involved. Visible light has a wavelength of around a few ten millionths of a meter or a few times 10^{-7} m, but to give a more specific example, light of a particular shade of green (the color to which the human eye is most sensitive) has a wavelength of 5.0×10^{-7} m. This is the same as 500×10^{-9} m, or 500 nm (a nanometer being equal to 10^{-9} m). Both physicists and astronomers use nanometers when talking about the wavelength of visible light, but astronomers also frequently use another unit, the *angstrom*, represented by the symbol "Å." One nanometer equals 10 Å, and so once again our green light can be said to have a wavelength of 500×10, or 5,000 Å. In more advanced books and research papers, you'll often find a wavelength written as $\lambda5,000$ or $\lambda6,563$, etc. When talking about a range of wavelengths in angstroms, this will be written, for example, as $\lambda\lambda5,000–7,000$.

At this point, it's worth also having a look at the kinds of numbers that are involved when we talk about the frequency of visible light. To do this, we rearrange Equation (1) in the form

$$\nu = \frac{c}{\lambda} \tag{2}$$

where we have substituted c (the speed of light) for V. The speed of light is very nearly equal to 3.0×10^8 m/s, and using 5.0×10^{-7} m for the wavelength, the frequency is given by

$$v = \frac{3.0 \times 10^8}{5.0 \times 10^{-7}} = 6.0 \times 10^{14} \tag{3}$$

This is a truly enormous number, which also illustrates why astronomers prefer to use wavelengths when dealing with the visible spectrum; 500 nm or 5,000 Å are much easier numbers to remember than 6.0×10^{14}. The frequency does have a very important role, though, as we'll see later.

Figure 4 shows the way that astronomers by convention divide the e-m spectrum up into its principal regions. The visible part of the e-m spectrum (often also referred to by astronomers as the optical spectrum) covers a wavelength range of about $\lambda\lambda 4{,}000$–$7{,}000$. As we head from $\lambda 7{,}000$ at the red end of the visible spectrum into the *infrared*, or "IR," region, we eventually reach a wavelength of $\lambda 10{,}000$ or 1,000 nm. One thousand nanometers is equal to 1 μm (short for micrometer) and so in the infrared region of the spectrum, particularly the far infrared, wavelengths are usually spoken of in microns – again, 2.3 μm, for example, is an easier number to deal with than 23,000 Å.

The infrared region is generally considered to extend to wavelengths of the order of a few hundred microns, whereupon we reach what astronomers refer to as the *submillimeter* region. This region takes us to a wavelength of 1,000 μm, or 1 mm, which by convention marks the short wavelength end of the *microwave* region, which covers wavelengths all the way up to 1 m. Your kitchen's microwave oven runs on a wavelength of around 12.24 cm; this wavelength, which incidentally corresponds to a frequency of 2.45×10^9, has a profound effect on water molecules – an effect that has not gone unnoticed by astronomers.

Wavelengths of 1 m and above make up the *radio* spectrum, and it's here that astronomers will often use frequencies rather than wavelengths. The radio region marks the low-frequency end of the e-m spectrum, but how low is low? Take for example a wavelength of 1,500 m, which is typical of traditional analogue long-wave radio broadcasts; the frequency is given by

$$v_{1500} = \frac{3 \times 10^8}{1.5 \times 10^3} = 2 \times 10^5 \tag{4}$$

for example, 200,000. Even at the low-frequency end, we have the electromagnetic equivalent of wiggling the end of a rope 200,000 times every second. One "electromagnetic wiggle" per second is given the name *1 hertz*, after the late 19th century German physicist Heinrich Rudolf Hertz.

So our 1,500 m radio wave has a frequency of 200,000 Hz, or 200 kHz. One important final point; if you ever need to convert a wavelength to a frequency and you use a value of 3.0×10^8 m/s for the speed of light, then *the wavelength must also be in meters* whatever part of the e-m spectrum you are dealing with. So, instead of using 6563 Å, use 6.563×10^{-7} m in Equation (2). Conversely, a frequency in hertz (which has units of "per second") combined with the speed of light in meters per second will give the wavelength in meters, which can then if one wishes be converted to angstroms or nanometers.

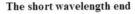

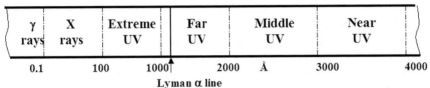

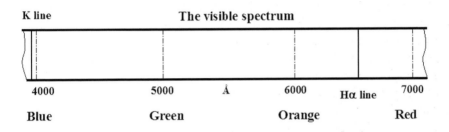

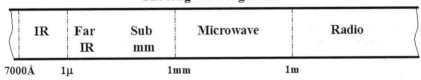

Figure 4. The principal regions of the e-m spectrum; wavelengths are not plotted to scale.

Heading the other way out of the visible spectrum we enter the *ultraviolet* region, which covers a wavelength range of λλ100–4,000. Medical practitioners and manufacturers of sunscreen products often make use of terms such as "UVA" and "UVB," which refer to specific regions of the ultraviolet, or UV, spectrum. Astronomers, however, divide this region into the near ultraviolet (λλ3,000–4,000), the middle ultraviolet

($\lambda\lambda$2,000–3,000), the far ultraviolet ($\lambda\lambda$1,220–2,000), and finally the extreme ultraviolet ($\lambda\lambda$100–1,210) – all pretty logical really except for those figures of λ1,210 and λ1,220. An important feature of the e-m spectrum due to hydrogen (the overwhelmingly most common element in the universe) is known as the Lyman-alpha line, which has a wavelength of about 1,215 Å, and this, by convention, marks the boundary between the two shortest wavelength regions of the UV spectrum.

We're all familiar with X-rays, usually within a medical context but over recent years astronomers, too, have become familiar with these very short wavelength (ranging from a mere 0.1 Å up to 100 Å) electromagnetic waves. Indeed, at wavelengths as short as these and with the even shorter wavelength gamma rays, it's customary for both astronomers and physicists not to use wavelengths at all but rather to talk in terms of the energy associated with the radiation.

We'll have a closer look at this shortly, but having briefly surveyed the electromagnetic spectrum, a good question to ask is this: How do you measure the wavelength of an electromagnetic wave in the first place?

How to Measure the Wavelength of Light

One way to measure the wavelength of light also gives us an opportunity to look at the classic experiment that showed that light was indeed a form of wave motion. The experiment was carried out shortly after the start of the 19th century by the English physician Thomas Young and is referred to as "Young's slit experiment."

The idea is to have two sources of the same kind of light next to each other. By "same kind of light" we mean two things; first, we need what is glibly called monochromatic light; that is, light of a single wavelength or color. We use the word "glibly" because, in reality, it's impossible to get light of a single wavelength. The best you can do is to have light that spans a very narrow range of wavelengths.

One way to do this would be to produce a spectrum by allowing sunlight to shine through a glass prism and then to block off all the light from the spectrum except for that which is allowed to shine through a very narrow slit. Even so, for Young's slit experiment to work, this slit has to be of the order of only a few hundredths of a millimeter wide.

The second thing is that the two light waves have to start from two points very close to each other and also they must start off "in sync," for example, if the peak of a wave crest comes out of one of the sources, then the same thing must happen at exactly the same time for the other source. The standard way to state this is to say that the two waves are *coherent* and are initially *in phase*. The way that this is done is to have two more narrow slits, close to each other and situated at the same distance from the monochromatic source slit. The monochromatic light will then pass through the two slits, and two coherent light waves will emerge in phase; then things get interesting.

A white screen is placed to intercept the light from the two slits. What's seen on the screen is a striking pattern consisting of an alternating regular series of light and dark bands. This happens because as the initially in-sync waves spread out from the two slits, there will be some locations between the slits and the screen where the waves get totally out of phase. In this situation the crest of one wave meets with a trough of the other wave, and the two simply cancel out – what's called "destructive interference." Conversely, there will be other locations where the waves are in phase, i.e., crest meets with crest or trough meets with trough, and we have "constructive interference." This situation also occurs at regular intervals along the length of the screen.

To produce a light band on the screen, the waves coming from the two slits have to be in phase, and for this to be so, the difference between the two distances traveled by the waves to the screen has to be a whole number of wavelengths. Conversely, to produce a dark band the waves must be out of phase, and the difference between the two distances has to be an odd number of half wavelengths.

Figure 5 shows the set up; clearly the two waves travel the same distance to the center of the screen and so will arrive in phase and produce a light band. For the light bands on either side, the waves have traveled distances

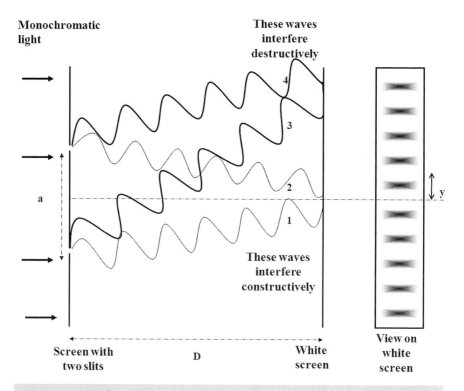

Figure 5. The principal of the Young's slit experiment. Waves such as those labeled "1" and "2" reach the white screen along paths whose distances differ by a whole number of wavelengths (in this instance the distances are the same); they thus arrive "in phase" and produce a bright fringe on the white screen. Waves such as those labeled "3" and "4" cover distances that differ by an odd number of half-wavelengths; they arrive "out of phase" and produce a dark fringe on the screen. The actual separation of the fringes is determined by the wavelength of the light and thus provides a means of determining the wavelength itself.

that differ by one whole wavelength, two whole wavelengths, and so on. Having previously measured with great care the distance "*a*" between the two slits and the distance "*D*" to the screen, the distance "*y*" between the central light band and one of its nearest neighbors is measured. This turns out to be given by

$$y = \lambda \times D/a \tag{5}$$

and so the wavelength of the light λ is given by

$$\lambda = a \times y/D \tag{6}$$

If you're good at trigonometry, you should have no difficulty in figuring this out; don't worry, however, if you're not. What really matters here is that the alternating pattern of light and dark bands, or *interference fringes* as they are often called, could only be explained by assuming that light was some form of wave motion. Equations 5 and 6 only serve to illustrate how straightforward it is to determine the wavelength of the light, by carefully making a few simple measurements.

Nowadays there are various ways to determine the wavelength of light, including theoretical calculations based on quantum mechanics. It goes without saying that astronomers need to be able to routinely measure wavelengths both in the visible and other parts of the e-m spectrum. However, as we shall now see, there is more to the visible spectrum than meets the eye.

The Spectrum, the Whole Spectrum, and Nothing but the Spectrum

In a laboratory a spectrum can be produced by passing sunlight through a glass prism. The resulting spectrum is the familiar continuous band of color ranging from violet through to red. Some of you may remember, though, being able to look through a *spectroscope* in the school physics lab at the spectrum produced by a hydrogen discharge lamp. This lamp consisted of a narrow glass tube containing hydrogen gas, which was connected to a high voltage power supply. You may also remember that when you looked through the spectroscope, you saw not a continuous rainbow of color but a series of individual bright lines set against an otherwise dark background. Hmm! Clearly not all spectra are the same. Furthermore, as you developed an interest in astronomy you will have soon come across images of the spectra of stars, which almost always showed the familiar continuous "rainbow" but this time crossed by dark lines. We shall now see where all of this came from.

The pioneering work of Robert Bunsen and Gustav Kirchhoff is generally regarded as the beginning of the science of spectroscopy as we know it. Indeed it was Bunsen and Kirchhoff themselves who invented the spectroscope – an instrument that could be used to produce and analyze the spectra from various sources of visible light. They discovered that there are three basic types of spectra, and in the following years, astronomers came to realize that all three have a role to play in the understanding of starlight. The three types of spectra are as follows:

- *The continuous spectrum.* This is the kind of spectrum we are all familiar with; a continuous band or "rainbow" of colors running from deep red through orange, yellow, etc., to violet. Bunsen and Kirchhoff found that this kind of spectrum is produced by something that is both very hot and dense. It has to be hot enough to be giving off visible light and dense enough, meaning that in Bunsen and Kirchhoff's time, suitable sources of a continuous spectrum would have been heated solid objects or perhaps molten metal.

- *The emission spectrum.* This spectrum consists of a series of discreet or individual bright lines (called *emission lines*) set against a dark background – just as you may have seen in the school physics lab. This type of spectrum, they found, was produced by hot, thin gas that they themselves produced by vaporizing various chemical substances in the hot flame of Bunsen's newly invented gas burner (the Bunsen burner). Their profoundly important discovery about this type of spectrum is that the pattern of lines is unique to the chemical elements that are contained in the thin gas. Vaporized copper salts, for example, give a completely different set of lines to those produced by iron or calcium salts.

- *The absorption spectrum.* The light from a very hot, dense substance is allowed to shine through cool gas (of low density) such as sodium or mercury vapor. This time the continuous spectrum is crossed by individual dark lines, and as with the emission lines the pattern of these *absorption lines* is unique to the chemical elements in the cool vapor.

In the following years, astronomers soon discovered that the light from most stars produces an absorption spectrum, though in some cases there are also emission lines superimposed on the absorption spectrum. Straightaway this told astronomers that stars consist of a source of a continuous spectrum, which must therefore be relatively dense as well as being hot. There must also be some cooler thin gas involved to produce the absorption lines, and some stars must also incorporate very hot, thin gas to produce emission lines. If this wasn't enough, the patterns of absorption lines in stellar spectra could be identified with identical patterns produced by known chemical elements in the laboratory, thus revealing for the first time not what starlight was made of but what the stars themselves were made of. And this turned out to be merely the tip of a truly enormous starlight "iceberg."

A very logical if rather obvious question at this point is why absorption or emission features are in the form of lines rather than, say, circular blobs. The answer is to many also obvious, but as someone once said, "These things have to be said anyway."

A Brief Note on the Spectroscope

The central feature of a spectroscope is the bit that produces the spectrum from the incoming light. In Bunsen's and Kirchhoff's day this was a glass prism; nowadays the same job is done much more efficiently by a device called a *diffraction grating*. Both devices enable light of a given wavelength to emerge traveling in a direction that depends on the wavelength itself. The result is that the range of wavelengths that make up a source of light are sequentially spread out to form a spectrum. This process is called *dispersion*, and the greater the dispersion the more the wavelengths are spread out. A diffraction grating produces spectra with greater dispersion than a prism, and this in turn enables finer details in the spectra to be observed.

Before the light reaches the prism or grating, it first passes through a narrow slit at the front end of a closed tube called a *collimator*. A lens at the other end of the collimator enables what is now the image of the illuminated slit to be ultimately viewed through a telescope (as in the high school spectroscope) or to be directly imaged with, say, a CCD camera. What enters the diffraction grating then is the image of the illuminated slit, which when dispersed by the grating, results in what amounts to a series of slit images spread out according to the wavelength of the light.

If we're dealing with an emission spectrum there will be several isolated slit images, one corresponding to each of the wavelengths in the spectrum. So each emission line is, in fact, the image of the spectroscope's collimator slit positioned according to wavelength. By contrast, a continuous spectrum is actually made up of a vast number of slit images all spread out according to wavelength and resulting in the rectangular band or strip of light that we are familiar with.

Finally, if light from some of the wavelengths in a continuous spectrum is removed by having cool gas in the way, then at these particular wavelengths no light at all is passing through the slit, so at these wavelengths the image of the slit is itself dark. So we have dark lines; i.e., absorption lines, crossing our continuous spectrum. In an absorption spectrum the continuous component upon which the absorption lines appear to be superimposed is always referred to as *the continuum*. Figure 6 summarizes the main results of Bunsen's and Kirchhoff's work.

It took a good half century and the development of quantum mechanics for a full and detailed explanation to be given of many of Bunsen's and Kirchhoff's results. This explanation started as a result of understanding the way in which light carries energy.

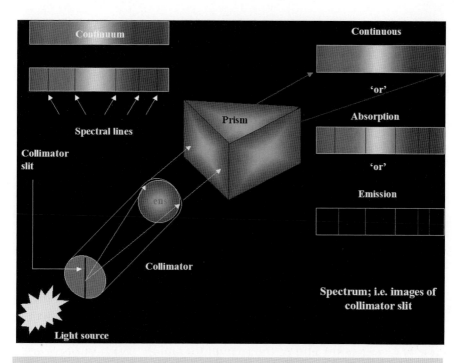

Figure 6. This summarizes the work of Bunsen and Kirchoff and shows why spectral lines are in fact in the shape of lines.

Light and Energy

One of the most important things about waves, including light waves, is that they enable energy to move around. When we wiggle the free end of our rope, our arm does work and puts energy in; a little while later and some meters further along the rope, the piece of rag moves up and down because it has received energy from the rope wave. Light waves, too, carry energy; sunlight can be used to generate electricity and also heat the ground, which heats the atmosphere and which in turn warms our planet.

In the classical electromagnetic wave theory of light, there are two possible ways to increase the amount of energy carried by the wave; one is to increase the maximum strength of the electric and magnetic fields along the wave, for example, to increase the amplitude of the wave by making the light brighter. This is analogous to making taller wiggles in our rope wave; we have to put more energy in to do this (our wiggling arm will start to ache sooner), and in consequence the piece of rag moves further because it gets more energy. Another way to put more energy into a rope wave is to wiggle the end of the rope more rapidly (again our arm tires more quickly); this means increasing the frequency of the wave, and again the same idea applies to an electromagnetic wave. So waves of higher frequency carry more energy, but as we've already seen and as Equation (1) tells us; a higher frequency means a shorter wavelength; the shorter the wavelength of light the more energy it carries. Thus, in the visible part of the spectrum, blue light carries more energy than red light.

Let There Be Another Idea About Light

If light consisted of Newton's energetic corpuscles rather than waves, then we'd expect that the two slits in Young's experiment would each produce a 12-bore shotgun-type spray of such corpuscles, which would reach the screen with a totally random distribution, producing a single band of light; only the wave theory could explain the alternating pattern of light and dark bands. Isaac Newton himself might well have been rather grumpy about the development of ideas on the nature of light during the 19th century; however, he would probably have had cause to smile around the turn of the 20th century, when it was found that in some situations the wave theory just didn't work. What's more, the theory that did work resulted from Newton's corpuscle idea being "reborn" – only this time the corpuscles were called *photons*. This more modern theory is called the *quantum theory of light*.

In our light wave thought experiments, we didn't say much about either the source or the detector of the light wave. Indeed, there was no need to, because we were concerned about the nature of the light wave itself, and in fact we could be forgiven for being lulled into getting the impression that our light wave was some sort of continuous steady thing, like the rope wave and water wave. The wave would eventually stop, of course, but only when we switched off the light. The source and the detector are important, though, because one actually produces the light and the other enables us to analyze it.

Both source and detector have one very important thing in common – they involve the interaction of light with matter. It was experiments involving this interaction, in one case a source of light, in the other a detector of light that turned out to be the downfall of the wave theory of light itself.

Without going too much into the details (we can do this later, when necessary), all matter is made of exceedingly tiny atoms that themselves are a (very intricate and highly organized) mixture of positive and negative electric charges. Electric charge is one of those fundamental things in nature, which as Mr. Taylor, the high school physics teacher, would probably have said, we'll never get to the bottom of.

Electric charge is the very source of electric fields, which as we've seen are an integral part of a light wave. The important thing here is that there are *two types of charge*, negative and positive. Even more important is the fact that, as they say; "likes" repel and "unlikes" attract, so negative charges are attracted to positive charges with what turns out to be a tremendously powerful force. In an atom, the positive charge sits in a tiny central nucleus that does, however, contain almost all of the mass of the atom. In a normal (or electrically "neutral") atom an equal amount of negative charge is carried by exceedingly low mass particles called electrons, which have their "place" on the periphery of the atom. The important point here, though, is not the structure of the atom itself but simply the fact that these electrons are bound to the atom by a mighty strong force. To overcome that force and remove an electron from an atom requires energy. But, as we've seen, light waves carry energy, so they can and do remove electrons from atoms. This wonderful phenomenon is called the *photoelectric effect*, and among many other things it makes an astronomer's CCD camera work.

Take some solid material and shine a light on it; for some materials this simple event will cause electrons to be removed from some of the atoms in the surface layers of the material. These electrons can in turn be led away along a wire in the form of an electric current that can be measured. Now let's see what happens when we decrease the amount of energy in the incoming beam of light. According to the wave theory, there are two ways we can do this: one is to decrease the amplitude of the wave, or simply making the light dimmer. The other way is to decrease the frequency of the light wave, for example, to increase the wavelength.

Let's lower the amplitude of the light waves first; as the light dims (provided we had an electric current coming out of the material in the first place), we'd find as we'd expect that the current gradually drops to zero as the light fades. So far, okay for the wave theory. Now let's lower the frequency (increase the wavelength) of the light; this time something totally unexpected happens. The current would not gradually drop to zero, but at some frequency it would suddenly stop. It's as though light below a certain frequency just hasn't got enough energy to shift those electrons. But surely, we say to ourselves, all we have to do is to turn up the brightness – increase the wave amplitude to compensate for the energy deficit caused by lowering the frequency. When we do this, however, nothing happens; the electric current doesn't come back on.

Our first thought might be that maybe the amplitude of a light wave is in some way tied into its frequency, so that lower frequency light waves have lower amplitudes. This might explain the lack of sufficient energy to remove an electron from the material's surface, but as a consequence, we would have modified the wave theory of light without any justification, save to account for the experimental result. This has all the hallmarks of introducing epicycles to explain the motion of the planets. If we think about it though, even a lower frequency wave with a low amplitude will surely after a time pump sufficient energy into an electron to shift it. The experiments, though, say "no"; it's like trying to make a hole in a hard brick wall with a blunt drill. No matter how long we drill for, we just don't get anywhere. The bottom line to all of this is that, as we said at the beginning of this chapter, theories of physics – in this case the wave theory of light – are models, and here we have a model that's not working. Time for a new model then!

Our photoelectric experiment suggests that what's needed to remove an electron from the material is a short, sharp kick rather than a steady push, and what's more, that kick has to be hard enough or the electron stays put. In fact, in order to make a new model of light, we have to suppose that this indeed is exactly what an electron gets from light; no steady push means that the idea of a continuous beam of light that consists of a continuous steady wave is gone. The short, sharp kick, rather than the steady push, means instead that light has to come in packages of energy (Newton's corpuscles if you wish, but the modern name for them, as we said, is photons). A photon has to have sufficient energy to knock an electron out of an atom, because in this new model, to quote an oft-used expression, the photon only gets one shot. It takes one photon *and only one photon* to remove one electron and *only one electron* from an atom. It is not possible to remove an electron by "pummeling" it with a rapid succession of lower energy photons.

Crucial to this new model then is the energy carried by a photon. Just on a fairly obvious point, all photons travel at the speed of light (c), just as with light waves, so speed isn't involved in this. According to our photoelectric effect experiment, the incoming light has to be at least a certain minimum frequency (as interpreted by the wave theory) in order to eject the electrons. The quantum theory effectively "steals" this very idea from the wave theory and uses it as if it were its very own. It gets away with it, too, because, as we've seen, determining the wavelength and hence the frequency of light is fairly straightforward. It makes the bold statement that the energy of a photon E is given by

$$E = h \times v \qquad (7)$$

The quantity "h" is a fundamental constant of nature called *Planck's constant*, and it has the value; 6.626×10^{-34} J s – the joule (as mentioned in the last chapter) being the standard unit of energy. See if you can work out the energy of a Lyman-alpha (1,220 Å) photon that has a frequency of 2.46×10^{15} Hz. (The answer is about 1.63×10^{-18} J.)

This new quantum theory of light doesn't even bother to try and reinterpret the idea of "frequency"; it simply uses it as a convenient way to calculate the energy of photons, and if you prefer to use wavelength rather than frequency, then that isn't a problem either, because Equation (2) tells us that

$$E = h \times c/\lambda \qquad (8)$$

So, based on the photoelectric effect experiment itself, the energy of a photon seems to depend on just one thing, and this is what is interpreted by the wave theory as the frequency (and by implication the wavelength) of the light. Indeed, experiments have shown that the energy of electrons that get kicked out in the photoelectric process is directly related to the frequency of the incoming light (provided the frequency is sufficiently high to remove them in the first place).

So we have a somewhat schizophrenic situation here; our new model of light says that light comes in packages, which are not waves but yet whose energy is proportional to the frequency that they would have if they were waves. On bad days this is the kind of thing that could make you feel like we live in an imperfect universe. However, in reality, it's more likely our theories of physics are imperfect.

One thing, though, that we should ask here is where the amplitude has gone. Photons don't have an "amplitude." Instead, when we make the light brighter we are simply producing more photons, and this results in more electrons being ejected; we observe an increased current. This is exactly the same effect as increasing the amplitude of a light wave. However, the energy of each individual photon is determined only by its frequency.

Summing up: one single photon with a high enough frequency and thus high enough energy can remove one electron from an atom; it takes many photons each with sufficient energy to remove many electrons from many atoms. Once you accept this (without worrying about exactly what photons are), you are able to understand a whole lot more astronomy than you could before, because as we shall see, the photoelectric effect is happening in some form or other all the time throughout the universe.

A good question to ask at this point is, what about those photons whose frequencies are too low to enable them to remove electrons from some material? Do they have any effect on the material itself? The answer is "yes," and the implications of this answer manifest themselves not only in the very character of the light from stars, but they also enable some of the fundamental properties of stars themselves to be determined.

The Parable of the Two Tables – Blackbody Radiation

It's a well known fact of everyday life that if you place something by a window in your house where the Sun shines in, the thing you've put there gets hot. This can of course sometimes have disastrous consequences, but it also tells us that the energy of a great many of the photons that fall upon any material is used to raise the material's temperature rather than to remove electrons.

Let's say you have a table, which just happens to be white in color, by the window in your living room; the table is bathed in the light (i.e., electromagnetic radiation) that is streaming in through the window. One thing that happens is that some of the radiation (particularly, in this case, radiation in the visible part of the spectrum) is reflected or scattered off the surfaces of the table, and some of this radiation can enter our eyes, which is how we are able to see the table in the first place.

Another way of "looking" at this is to note that this reflected or scattered radiation is radiation that the table has failed to absorb; in other words, our table has done a less than perfect job at being an absorber of radiation. Some of the sunlight, though, will be absorbed by the table, and it is this absorbed radiation that causes the table's temperature to rise. The most important consequence of this is that the table will emit its own radiation, besides absorbing and scattering incoming sunlight. Indeed, any object that is at a temperature above absolute zero (this is zero degrees, 0 K on the Kelvin temperature scale, or −273°C) will emit electromagnetic radiation; this radiation comes from many atomic and molecular processes going on within the object that are directly related to the object's temperature, and so it is called *thermal radiation*. This means, of course, that even before the Sun shone through the window, the table was already emitting thermal radiation. The effect of absorbing the incoming sunlight is to raise the temperature of the table still further, which results in increased emission of thermal radiation.

Now suppose that we have a special table; this new table absorbs *all* of the radiation that falls on it, i.e., all colors and all wavelengths. Everything is absorbed by this special table. Because no light is being reflected or scattered, the table would now appear as a perhaps rather sinister black shape with the outline of our table. Just as with the ordinary table, the temperature of this special table will rise as it absorbs sunlight; however, because this special table absorbs all of the incoming radiation the temperature rise will inevitably be greater than it was for the ordinary table.

The result is that the special table will emit more thermal radiation than the ordinary one. The two tables are shown in Fig. 7.

Figure 7. Here we see two tables; the one on the left is an "ordinary" everyday table, which just happens to be colored *white*. Its spectrum will consist partly of an infrared thermal spectrum together with some reflected and scattered visible radiation. By contrast the table on the right is a "special" blackbody table. It absorbs all radiation falling on it and emits a purely thermal spectrum that, at room temperature, will be largely in the infrared region.

The first thing to say about the thermal radiation emitted by both tables is that at the kind of temperatures that pervade the average domestic living room, this radiation will be in the infrared part of the e-m spectrum. A very important question here is why the ordinary table absorbs some sunlight, but some visible light in particular is reflected or scattered, giving the table its white color. The special table, however, absorbs this visible light but effectively re-emits it as infrared radiation. So how did the visible light get changed into infrared radiation?

In a very general way (without going into a detailed discussion of atomic and molecular absorption and emission processes, which are, in fact, described in more detail in the author's *Spectroscopy: The Key To The Stars*, also published by Springer), it's possible for an atom to absorb, say, a blue photon; this is a fairly high-energy photon for the visible part of the spectrum. In the quantum theory there is just a possibility that this atom will then emit an identical blue photon, effectively re-emitting the photon that it absorbed. However, there is an overwhelmingly greater probability that the atom, which for a very brief moment has extra energy by virtue of having absorbed the blue photon, will lose this energy in the form of a short "machine gun burst" of lower energy, i.e., infrared photons. In fact, this process of higher energy photons being degraded into lower energy photons is something that tends to always happen when light interacts with matter.

Just a final note here. The chance of an atom absorbing two or more photons in succession and thus building up an even larger store of extra energy is even more remote, because a typical atom only holds onto its extra energy for a period of around a one hundred millionth of a second. This feature of the quantum theory enabled it to successfully explain the spectrum of thermal radiation from hot bodies. The wave theory of light fell down on this, because it assumed that it was possible for an atom to absorb energy continuously from an incoming electromagnetic wave. This energy could then be re-emitted in the form of a very high-energy electromagnetic wave such as ultraviolet light. This would mean, for example, that we could get a suntan from our table just by placing it in the window. This does not happen, so clearly the wave theory is wrong here.

So we have two tables bathed in sunlight and doing slightly different things. The way to investigate further the difference between the tables is to consider the spectrum of electromagnetic radiation running from the infrared through the visible region, which is coming from them. Our tables are fairly dense objects and so we expect this to be a continuous spectrum. From the ordinary table some of the radiation has been reflected or scattered from its surface, and this we would expect to produce a peak or a "hump" in the visible part of the spectrum. The remaining radiation is thermal, and this would produce a hump in the infrared region. For the special table, all of the radiation coming off is thermal radiation, and so there would just be an infrared hump. However we would expect the infrared hump to be bigger than that for the ordinary table, because the special table has acquired that extra energy that was scattered by the ordinary table.

Another way to look at this is to say that the thermal radiation from the ordinary table is deficient, because some of the incoming radiation has been lost by being scattered. In fact all "ordinary tables" or all "ordinary objects" for that matter will have this deficiency in their thermal radiation spectrum, compared to that of our special table. This table produces the most complete thermal radiation spectrum possible, simply because it has absorbed all of the radiation that falls on it. Objects like this are called *blackbodies* because (as indicated above) they would appear utterly black at ordinary room temperatures. The thermal radiation emitted by blackbodies is, not surprisingly, called *blackbody radiation*.

Here's another question: How hot will our tables get? Indeed, could it be that their temperature will continue to rise for as long as we leave them there by the window – or at least until the Sun goes down? The answer to this one at least is, of course, no; the tables will reach a stage when the rate at which they lose energy by thermal radiation exactly matches the rate at which they absorb it from the sunlight. A steady state will have been reached, which is called *thermal equilibrium* (this is often denoted in the literature simply as "TE"); for our ordinary table, as indeed is the case for any "ordinary object," we would have to carry out regular temperature checks in order to find out when thermal equilibrium had been reached by noting that the temperature remained steady. What's more, the actual temperature at which thermal equilibrium occurred would be different for every object, simply because different objects or different materials scatter – or, more to the point, fail to absorb electromagnetic radiation in different ways. By contrast our blackbody table, or for that matter any blackbody, has absorbed all of the radiation, or, in other words, all of the energy that has fallen on it, so that at thermal equilibrium the total amount of energy that is radiated each second by each square meter of the blackbody's surface effectively equals the rate at which energy is being absorbed by each square meter of its surface.

Assuming the Sun's energy output doesn't change this means that we could place several blackbodies in our window and at thermal equilibrium the rate at which energy is absorbed by each square meter of their respective surfaces would be the same. The result is that they would all have the same temperature and would emit the same amount of energy (E) each second from each and every square meter of their surfaces. A higher rate of absorption of energy would result in a higher temperature and a correspondingly higher rate of radiation of energy. Thus, for any blackbody the rate at which energy is radiated from each square meter of its surface is directly related to its temperature and is given very simply by

$$E = 5.670 \times 10^{-8} \times T^4 \mathrm{W} \qquad (9)$$

T is the temperature in degrees Kelvin, and the number 5.670×10^{-8} is called *Stefan's constant* – usually represented in the literature by the lower case Greek letter sigma, or "σ". This very simple formula is called the *Stefan–Boltzmann equation*, after two 19th century Austrian physicists – Josef Stefan, who worked out the formula by doing experiments on the way hot bodies cool down, and Ludwig Boltzmann, who derived it from theoretical principles. Simple as it is, the Stefan–Boltzmann equation is, as we shall see later, of huge importance for stellar astrophysics.

The Stefan–Boltzmann equation tells us two things. First, it tells us that the rate at which energy is emitted from each square meter of the surface of a blackbody *depends only on the body's temperature* and is in fact proportional to the fourth power of its temperature. Second, it tells us that the actual value of this temperature is determined by the blackbody being in thermal equilibrium with its local environment, i.e., the blackbody's rate of emission of energy equals its rate of absorption.

What this means is that to increase the temperature of a blackbody, we need to increase the rate at which it absorbs energy. One way to do this for our blackbody table would be to move it closer to the equator so that the Sun is closer to the zenith; this results in the sunlight falling on the table being more concentrated, resulting in the table getting hotter. Next, we could take the table out into space and physically move it closer and closer to the Sun; again this would result in more energy falling on each square meter of the table's surface with a subsequent rise in temperature. So each time we got closer to the Sun, the table's temperature would certainly increase, but it would then level off as thermal equilibrium was reached. Even if our table were made of some indestructible material, as yet unknown to science, and we moved it right up next to the Sun, the same thing would happen; the table's temperature would increase and then level off as thermal equilibrium was reached. Even our mighty Sun has its limitations when it comes to raising the temperature of a blackbody.

There is however another way to increase the temperature of a blackbody, and this is to change the *character* of the radiation falling on it; this means having radiation that contains more high-energy photons such as ultraviolet and maybe even X-ray photons. This will certainly increase the rate at which energy is absorbed by the blackbody and will result in a corresponding significant rise in the temperature, accompanied by an increase in the rate of emission at thermal equilibrium.

If changing the character of the spectrum of radiation absorbed by a blackbody will change its temperature, the next question then is, how does changing the temperature of a blackbody affect the spectrum of thermal radiation that it emits?

The Blackbody Spectrum

We've already suggested that blackbodies must be relatively dense objects (i.e., they would consist of solid or liquid material or a dense gas). As we recall from Bunsen's and Kirchhoff's experiments, a thin gas could never act like a blackbody because it will only absorb or emit photons at selected wavelengths. So the spectrum of thermal radiation from a blackbody – a *black body spectrum* – will consist only of a continuum with no absorption or emission lines. Because a blackbody spectrum does not necessarily involve visible light, it is standard practice to represent it by means of a graph that plots energy emitted against wavelength.

As already mentioned, at room temperatures most of the thermal radiation from a blackbody is in the infrared region of the spectrum. One thing that is very clear is that if we raise the temperature, the overall level of emitted thermal radiation will also rise, and so the blackbody spectrum will move further up the energy scale. The other really important feature of a blackbody spectrum is the wavelength at which the largest amount of energy is being emitted. If we adopt our "brute force" method of raising the temperature of a blackbody by moving it closer to the Sun, there will be an overall increase in the number of solar photons absorbed by the blackbody, and this includes an increased number of higher energy photons. Remember, we said that the chance of higher energy photons being re-emitted was very small, due to the overall degrading of absorbed photons; however, an increased number of higher energy photons absorbed means that there will also be a corresponding increase in the number of emitted photons, which themselves have higher energies. Our more subtle approach to raising the blackbody's temperature by placing it in the presence of radiation, which contains very high-energy photons, will certainly produce more thermally emitted high-energy photons. The result is that as the temperature increases the wavelength at which maximum emission occurs will become shorter.

This process is "seen" (probably by most of us on TV at some time) when a piece of steel is heated; at first it simply gets hot and emits most of its thermal radiation in the infrared. As the temperature rises, the steel begins to glow a dull red. This shows that the peak emission wavelength has moved into the visible part of the spectrum. With further increase in temperature the steel glows bright orange and then white, and finally it may acquire a hint of blue as the peak emission shifts to even shorter wavelengths. At this point the significance of the colors of stars becomes very obvious.

There is in fact a very simple formula (remember, though, that this formula only works for blackbodies) connecting the wavelength λ_{max} of this peak emission with the temperature:

$$\lambda_{max} = 2.8973 \times 10^7/T \tag{10}$$

The wavelength λ_{max} here is in angstroms, and the temperature is in Kelvin. This formula is known as *Wien's law*, or sometimes Wien's displacement law (pronounced "veen" and named after the late 19th century/early 20th century German physicist Wilhelm Wien). Try working out a few maximum emission wavelengths using your calculator by plugging different temperatures into the formula; for example, a temperature of 5,800 K – the surface temperature of the Sun – gives a λ_{max} of 5,000 Å in the green region of the spectrum.

The actual shape of a blackbody spectrum is that of an asymmetrical "hump"; the slope of the curve is steeper on the short-wavelength side than on the long-wavelength side. As the temperature of the blackbody rises, the spectrum moves further up the energy axis, and the peak of the curve progressively moves to shorter wavelengths. Figure 8 shows a series of blackbody spectra plotted for a range of temperatures.

The German physicist Max Planck was the first person to derive a theoretical formula for the blackbody spectrum. However, his formula is rather complicated, and we ourselves (fortunately) don't really need to use it. Much more important for our purposes is the relatively simple Stefan–Boltzmann equation and the Wien's displacement law.

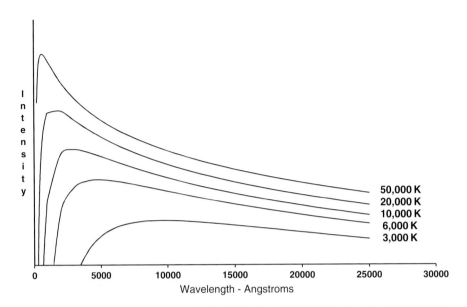

Figure 8. These curves plot the emitted intensity of blackbody radiation against wavelength. Notice first that as temperature increases, so does the amount of emitted radiation across all wavelengths. Second, the plots show very clearly how the wavelength at which maximum emission occurs decreases with increasing temperature. Notice finally the asymmetrical shape of the curves, i.e., a sharp rise on the short-wavelength side but a much more gradually sloping long wavelength "tail."

The Parable of the Blackbody and the Star

Stars do not radiate as blackbodies; their spectra usually consist of a continuum together with absorption lines and sometimes emission lines. The continuum part of a star's spectrum is, however, similar to a blackbody spectrum – at least to a first approximation. This is a very fortunate situation for astronomers, because the properties of blackbodies are well understood; it means, for example, that it is possible to make an estimate of a star's temperature without having to send a space probe there with an attached thermometer.

On a more subtle level, the way in which a star's spectrum actually differs from that of a blackbody can probably tell us more about the star than anything else. We shall begin this subtle journey on the river of starlight by asking a simple and what at first might seem like a trivial question: Just how bright is a star? The answer will actually take up the next two chapters, but at the end we will be able to "take a star's temperature." Figure 9 shows some typical stellar spectra that clearly resemble the blackbody curves of Fig. 8.

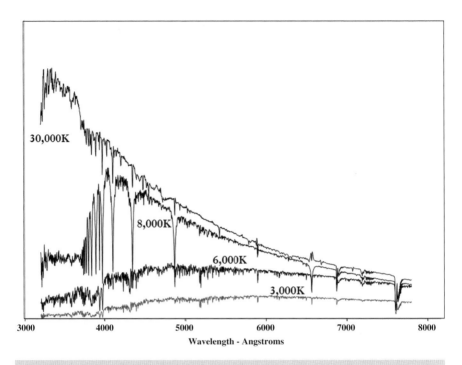

Figure 9. Typical spectra of stars that have a range of temperatures. Notice their obvious resemblance to blackbody spectra of corresponding temperatures. (Spectra reproduced from the STELIB library of stellar spectra by kind permission of Jean-François Leborgne at the Laboratoire d'Astrophysique de Toulouse Tarbes, Université de Toulouse.)

Key Points

- In classical (pre-20th century) physics, light was thought to consist of electromagnetic waves.

- The wavelength of an electromagnetic wave determines the color of the light or the character (infrared, ultraviolet, etc.) of other forms of electromagnetic radiation.

- Electromagnetic radiation carries energy.

- The quantum theory assumes that electromagnetic radiation comes in the form of packets of energy called photons.

- The energy carried by a photon is directly related to its classical frequency.

- All bodies with a temperature above absolute zero emit thermal radiation.

- A perfect absorber of radiation is called a blackbody, which in turn emits the most complete spectrum of thermal radiation.

- The continuum part of the spectra of stars resembles a blackbody spectrum to a first approximation.

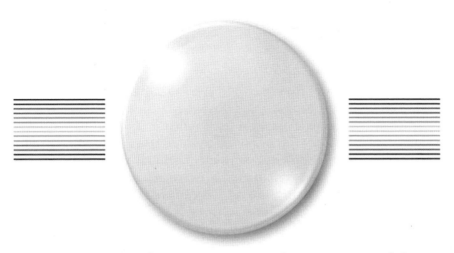

Space – The Great Radiation Field

Space is not empty! Well of course these days we know that it seems to be full of dark matter and dark energy, but these (can I say?) exotic things aside, the space between galaxies, between stars, and even between the particles, molecules, and atoms of the interstellar medium contains an endless river of radiation – electromagnetic radiation, which in this chapter we'll simply call "light." What's more, aside from the cosmic microwave background, most of this radiation began in stars, and some of it will reach our eyes, our telescopes, and other instruments. As astronomers we can then use this precious stuff to learn much of what we know about the stars themselves, their immediate environments, and even what's going on in the space between the stars. In this chapter we'll investigate the flow of starlight across space and see that in the end it manifests itself to us as something that is familiar to all astronomers – the magnitude scale, which, as we'll see, is a truly elegant way of measuring the energy that we receive from the stars.

K. Robinson, *Starlight*, Patrick Moore's Practical Astronomy Series,
DOI 10.1007/978-1-4419-0708-0_4, © Springer Science+Business Media, LLC 2009

Rivers of Energy

Earth is surrounded by stars, which means that a river of starlight is constantly flowing past us from each and every one of those stars. Physicists tell us that light carries energy, and this in turn tells us perhaps the single most important thing about light itself – that it is one of the principal agents by which energy is transported through space and across the universe. In any region of space these rivers of light taken together make up what astronomers call a *radiation field*. So when you observe with your telescope either visually or using a CCD camera to do maybe photometry, spectroscopy, or to produce deep sky images, you are in fact sampling and in effect measuring some of the energy contained within the radiation field in our part of space.

Measuring the Radiation Field

Physicists are always doing "experiments" in their heads; they call these experiments "thought experiments." As amateur astronomers we, too, can carry out thought experiments, but these usually need to be "done" in space, where Earth's atmosphere won't complicate things. So let's kit ourselves out as thought experimental astronomers. Aside from the proverbial spacesuit, we will need some sort of detector, like maybe a very advanced CCD camera that can receive and measure energy from the surrounding radiation field. This camera has a larger than average CCD chip, which measures 1 m by 1 m, i.e., a surface area of 1 m^2. In the vicinity of Earth the surrounding radiation field is overwhelmingly dominated by one source – the Sun. In order to understand better what a radiation field is all about, let's put several light years between the Sun and ourselves, enough distance so that the Sun looks basically just like all the other stars that we can see. Now we have a radiation field that results from a more even distribution of sources; namely all the stars, galaxies, nebulae, etc., that surround us. This radiation field consists of a flow of light, that is, a flow of energy coming from many different directions – but not all directions.

One of the great problems of cosmology was the so called "Olbers' paradox," which said that if the apparent gaps between the stars were in fact occupied by yet more distant stars, then their light, too, would eventually reach us, and so the space all around us would appear blazingly bright. This is not the case, of course, so there must effectively be directional gaps in the radiation field. We can see this if we point our super CCD camera directly at a star – say a distant sun; the detector can register and measure the amount of energy it receives. Now if we point the camera slightly away from that sun, the reading will fall off. This pattern will repeat itself as we move our camera from star to gap between stars, to star, etc. So clearly the radiation field is not the same in all directions.

If, on the other hand, we carried out this experiment from the center of a globular star cluster such as Omega Centauri or M13, then we would find it harder to detect those dips in the levels of the radiation field, because the apparent gaps between the stars would be smaller. Here the radiation field would be much more uniform in all directions. A radiation field that *is* totally uniform in all directions is said to be *isotropic*. This is a hypothetical idealized radiation field; in reality, all radiation fields in space are non-isotropic, or *anisotropic* – even the microwave background is anisotropic.

The size of our detector's chip means that when we place it in a radiation field consisting, for example, of the light from several distant stars, the detector measures the total amount of energy that crosses an area of 1 m². This energy, though, is something that is flowing constantly; every second, our detector receives a quantity of energy that we can measure. As mentioned previously, we measure energy in "joules," named after the English physicist James Prescott Joule. How many joules our detector receives in 1 s is measured in another unit; the "watt," named this time after the Scottish engineer James Watt. The watt is the basic unit of *power*; so power is the rate at which energy is either used or, in our case, received by the detector. So when we use our detector to measure the surrounding radiation field, we are in fact measuring the power which we receive from distant stars. Astronomers actually use a different word for this power in the radiation field; they call it the *flux* or sometimes the *radiant or radiative flux*, and this comes in units of so many *watts per square meter*.

The radiation field contains energy that is constantly flowing. If we imagine a transparent box with a volume of 1 m³, at any given instant of time our box will contain a certain amount of energy. In the next instant this will be replaced by fresh energy from the flow in the radiation field. If, while we're doing our thought experiments, nothing unusual happens (no supernovae, for example, and let's for the moment forget about all those variable stars out there), then we can reasonably expect our one cubic meter box to contain the same amount of energy from one moment to the next.

If we now move our box around to different locations in our region of space and we find that wherever we go the box always contains the same amount of energy at any given time, then we can come to the grandiose conclusion that the radiation field in our region of space is *homogeneous*. The notion of a homogeneous radiation field is again an idealization, just like the idea of an isotropic radiation field. In reality, of course, there *are* those variable stars out there, and from time to time a nova or even a supernova would outburst. These things constantly change the energy density of the radiation field and also modify its isotropy (or strictly its anisotropy). These constant changes – large or small – rapid or leisurely in the radiation field surrounding Earth are what make life so interesting for the astronomer.

Now we aim our detector directly at the distant sun and measure the flux, i.e., the number of watts per square meter that we receive from it. Notice we've used the word "directly" here; this means that the incoming radiation from the sun arrives at the detector in a direction that is at right angles to the plane of the detector. Another way to phrase this (used extensively in the literature) is to say that the flow of radiation is *normal* to the plane of the detector. Of course, radiation from many other stars, etc., would also be falling on the detector – after all, when you observe with a telescope either visually or with a camera, you would generally expect to see more than just one single star in the center of your field of view. Because our detector isn't pointing directly at these other sources, the flow of radiation from them is not normal to the plane of the detector. The detector in fact presents a cross-sectional area of less than $1\,\text{m}^2$ in these other directions, and the further we move away from the line that is normal to the plane of the detector, the smaller the cross-sectional area will become, as shown in Fig. 1. This results in the detector receiving less flux from these peripheral sources.

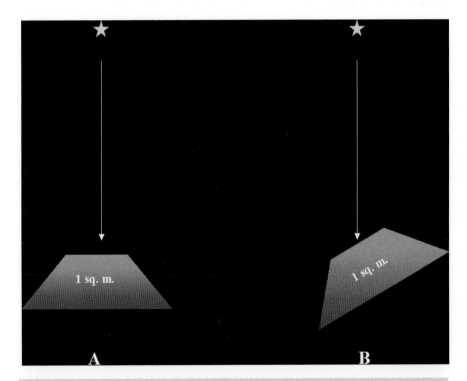

Figure 1. The plane of detector "A" is "normal" to the direction of the light from the star and so it receives the maximum possible flux, whereas detector "B" presents a smaller cross-sectional area to the direction of the incoming starlight and so measures a lower flux reading.

Nonetheless we add up the contributions from all sources to give us the total flux received by our 1-square meter detector. Now we turn the detector about face and do the same kind of observation to determine the total number of watts per square meter coming from the opposite range of directions. We now have two numbers, measured in watts per square meter; so now we subtract the smaller number from the larger one to give the *net rate of flow of energy across one square meter of the radiation field*, as shown in Fig. 2. This is called the *total radiative flux*, or just simply the *total net flux* through this point in the radiation field, and as you'd expect it's measured in watts per square meter.

Finally, if we confine our attention to one single object, such as a distant star, then clearly the flux from this one source involves radiation that is effectively confined to one direction. (In reality, it is confined to a very narrow range of directions spanned by the diameter of our detector.)

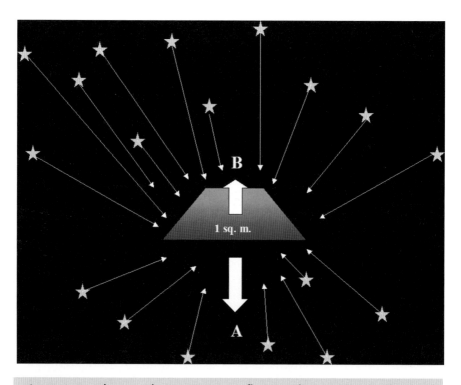

Figure 2. Flux reading "A" minus flux reading "B" gives the total net flux (per square meter) through a given point in the radiation field. For a radiation field homogeneous and isotropic, the two readings "A" and "B" are equal, and the total net flux through any point is zero.

In the idealized radiation field, which is homogeneous and isotropic, the net flux through any region of space would be zero, and what's more, this situation would never change. This would make for a very dull universe; fortunately, though, as we've said, real radiation fields are anisotropic, and the net flux through most any region of space is not zero. The light from any individual star, for example, would flow in one direction through a 1 m^2 area placed in its path. Close to the photosphere of the star, the anisotropy of its radiation field would be greater still, and even down inside the star there is a net outward flow of radiation. In fact, it's probably not too great a simplification to say that all of astronomy comes down to knowing how many watts per square meter flow past a particular point in space.

Sunlight Is Intense – But Starlight Is Not

Okay, we've learned that the radiation field is made up of a constant flow of energy coming in many directions from distant stars' galaxies and nebulae. Let's return now to the neighborhood of Earth and examine the radiation field in our part of space. Of course the radiation field here is overwhelmingly dominated by the flow of energy from the Sun; our CCD detector will receive a much greater level of flux from this source than from the distant stars. There is, however, something else about the Sun that is different from the other stars. The Sun appears as a disk rather than as a point of light. This is a very obvious observation, but it introduces to us a subtle and important feature of the radiation field.

Because the Sun has an apparent angular size as seen from Earth (about half a degree), we are able to observe that some parts of the visible solar disk, or the photosphere, appear darker than others. Sunspots are distinctly darker, and the region around the edge or limb appears distinctly dimmer than the central regions. All of this leads us to conclude that the flux that emerges is different for these different regions.

If we take a very small area of the photosphere, though – 1 m^2 is itself very tiny compared to the size of the Sun – then we can be fairly safe in assuming that the emergent radiation is uniform across this tiny region. This radiation will spread outward in all directions – indeed, some of it will head back down into the deeper layers of the Sun. However, we cannot assume that the amount of radiation that is heading outward will be the same in all directions. So, in addition to starting with a tiny surface area, we also need to take a tiny range of directions so that we can be sure that the amount of radiation is uniform across this range of directions. The range of directions that we are most interested in, of course, is that which includes "our direction."

A "range of directions" heading outward from a tiny surface area will actually form a cone-shaped region (actually, a slightly truncated cone), with the surface area itself forming the "point" of the cone, as shown in Fig. 3. Now imagine this cone extending outward from the Sun's photosphere to very large distances; what we can be sure of is that the total quantity of radiation that is confined within the cone and which passes some point in it in 1 s is the same *no matter how far away we are from the photosphere itself.* Finally, a narrower cone means less radiation passing through per second than a slightly fatter cone. What defines the narrowness of the cone – in other words, the range of directions – is the cone's

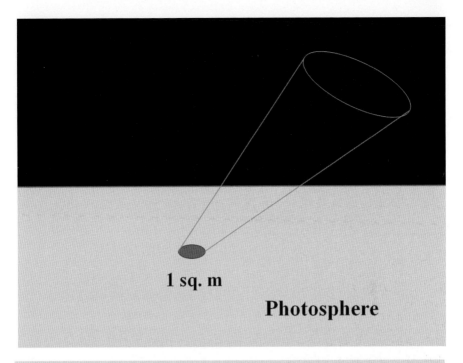

1 sq. m

Photosphere

Figure 3. The radiation emitted from a small unit surface area of the Sun's photosphere and which is confined to a tiny range of directions forms a cone shaped region.

so-called opening angle. This is formed by revolving an isosceles triangle about the angle at its apex, and it results in what mathematicians call a *solid angle* (in case you're interested, it's measured in what are called *steradians* and not degrees).

If we divide the quantity of radiation by the opening angle of the cone itself we have the amount of radiation passing per second per unit solid angle or, in effect, per "unit range of directions." This is called the *intensity* or sometimes the *specific intensity* of the radiation field. It is measured in watts per square meter per unit solid angle, and as we have seen it does not change with distance. What does change with distance, of course, is the cone's cross-sectional area, which increases, and this means that the quantity of radiation passing through a unit cross-sectional area decreases. This is our old friend the flux. If we can catch and measure all of the radiation confined to this range of directions then we are measuring the intensity; otherwise, we are measuring the flux, as shown in Fig. 4. This is the case with virtually all distant stars.

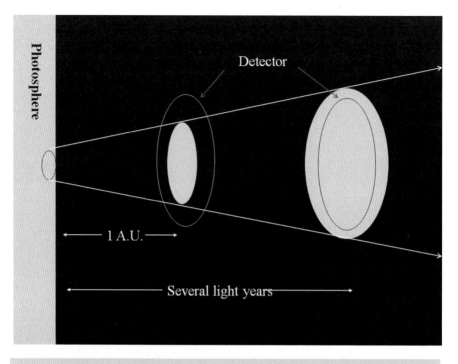

Figure 4. If all the radiation confined within a "unit range of directions" can be collected then it is possible to measure the intensity. With increasing distance the cross-sectional area of the cone shaped region increases until it is only possible to measure the flux. This is the case with most stars.

How Many Light Bulbs Must You Screw in to Make a Star?

The power rating of a domestic light bulb is given in watts – maybe 100 W. If we screw the light bulb into the socket in a modest apartment room and switch it on, the light spreads out in all directions (ignoring the obstruction caused by the bit that screws in). The room now contains a radiation field just like space; the light will reflect off the walls and furniture, enabling us to see things. If we now screw in our 100-W light bulb at the center of an aircraft hanger, we will see immediately that by the time the light reaches the much more distant walls, the radiation field has become very weak indeed, resulting in us being able to see very little. Clearly there is a lesson to be learned here about starlight, if we think of stars as huge light bulbs – and that's just what astronomers do!

A star such as the Sun gives out energy that spreads outward in all directions into the surrounding space. The rate at which the star is pumping out energy gives the power rating of the star; it's measured in watts, just like a light bulb, and it's called the star's *luminosity*. The luminosity of the Sun is a staggering 3.827×10^{26} W, or nearly 400 trillion trillion watts. So we'd need to screw in about 4 trillion trillion light bulbs to make a "star" like the Sun. In the literature, luminosity is invariably represented by a capital "L," and this in itself would be given in watts. However, more often than not, the luminosity of a star is given as so many times the luminosity of the Sun, which is itself denoted by "$L_\odot$" ("$\odot$" being the standard astronomical symbol for the Sun, of course). So, for example, the luminosity of a star that is 10 times that of the Sun is denoted by $10L_\odot$; one with 0.6 of the Sun's luminosity is denoted $0.6L_\odot$, and so on.

The ideal way to take advantage of and make use of this enormous amount of power flowing from a star would be to construct a sphere that totally enclosed the star and then live on the inside surface of the sphere, as shown in Fig. 5. This idea was conceived by the English astronomer Freeman Dyson, and such a sphere is, not surprisingly, called a "Dyson sphere." The idea was used to great effect in an episode of "Star Trek – The Next Generation," back in the early 1990s. Here the Dyson sphere gives us a way to carry out another thought experiment, to explore what happens to a star's energy as it spreads out into space.

However big we make our Dyson sphere (whatever its diameter), one thing's for sure – the inside surface of the sphere will capture the entire energy output of the star, which itself is simply a measure of the star's luminosity. This energy will be spread around the inner surface of the sphere, and the distribution of the energy will be completely uniform

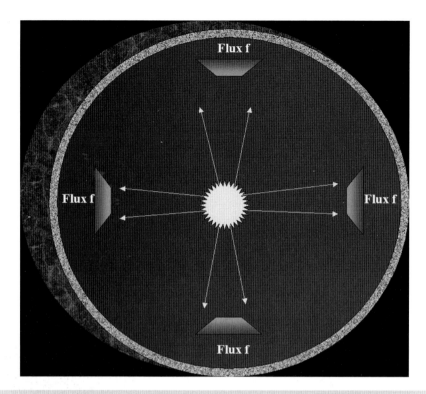

Figure 5. A Dyson sphere could capture the entire luminous output of a star. The flux received over every square meter of the sphere's inner surface would be the same, and this would be equal to the star's luminosity divided by the sphere's surface area.

(provided the star doesn't undergo any kind of eruption, such as a coronal mass ejection, which might send an extra boost of energy in some preferred direction). The amount of energy received each second by each square meter of the Dyson sphere's surface (i.e., the flux) will then be equal to the star's luminosity, divided by the sphere's surface area.

If we now dismantle this Dyson sphere and construct a bigger one, we still capture all of the energy from the star, but this time the energy is spread out over a bigger surface area. So the flux received by each square meter will now be smaller. The surface of this new bigger Dyson sphere is further away from the star than the previous one, so clearly the flux we measure depends not just on the star's luminosity but also on how far away from the star we are.

The surface area of our Dyson sphere or any sphere of radius "r" for that matter is equal to $4\pi \times r^2$ ("π" is, of course, the important number from high school geometry and is equal to 3.142 to three decimal places), and if we denote the luminosity of our star by "L," then the flux "f" at the surface of the sphere is simply equal to

$$f = \frac{L}{4\pi \times r^2} \tag{1}$$

If we double the sphere's radius, the surface area multiplies by four (2×2); if we triple the radius, it multiplies by 9 (3×3), and so on. So doubling the radius of the sphere would reduce the flux from the star to one quarter, and tripling it would reduce the flux to one ninth of its original value. The flux, in fact, follows an inverse square law similar to that for gravity, but here it's due to the way the surface area of a sphere increases as we increase the radius.

Now let's take away the Dyson sphere (it's still just science fiction at this stage anyway). We're out in free space again, as is represented in Fig. 6, but the same rules still apply; the flux that we receive from a star depends directly on its luminosity and inversely on its distance squared.

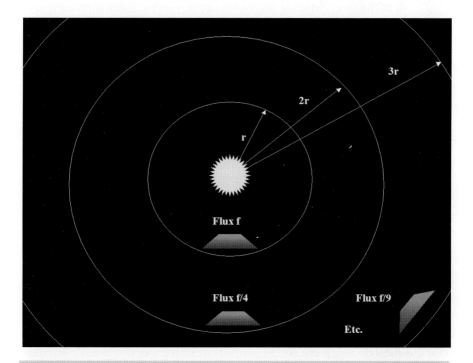

Figure 6. The flux received from a star decreases as we move further away from the star, and its value is directly proportional to the reciprocal of the star's distance squared.

Let's try to get a "feel" for how many watts per square meter, i.e., the flux that we might expect to get from a star. The best place to start is of course with the Sun, which as we know is pumping out energy at a rate of 3.827×10^{26} W. Here on Earth our average distance from the Sun is about 1.5×10^{11} m s (1 A.U.), and so a sphere of radius equal to this distance would have a surface area of $4\pi \times (1.5 \times 10^{11})^2$ m^2 s. This equals approximately 2.8×10^{23} m^2. So to get the flux we need to divide the Sun's luminosity by this area, and this gives us about 1.4×10^3 W or 1.4 kW/m^2. This value resulting from a slightly approximated calculation is itself approximately equal to a very important number – *the solar constant*. This is equal to the amount of flux passing *normally* (i.e., at an angle of 90°) through 1 m^2 at the top of Earth's atmosphere, and its officially accepted value is 1.366 kW/m^2.

Now let's "move" the Sun to a distance of 32.6 light years or 10 parsecs (pc); this amounts to a distance of about 3.1×10^{17} m s. A sphere of radius equal to this distance would have a surface area of 1.2×10^{36} m^2, and once again dividing the Sun's luminosity by this figure results in a flux of about 3×10^{-10} W/m^2. This very small figure gives us some idea of how much energy we might expect to receive from a modest star, which is situated at a modest distance from Earth. A star of greater luminosity at this distance will produce a greater flux value on Earth and will appear brighter both to our eyes and our instruments. In other words, *the apparent brightness of a star depends on the amount of flux we receive from it*, and this, in turn, depends on both the star's luminosity and its distance.

The word "brightness," though very intuitive and much used when discussing stars (as we shall indeed do here, where we regard "brightness" as amounting to the same thing as "flux value") is nonetheless somewhat vague. Flux, on the other hand, is a precisely defined quantity; it is the amount of energy that flows every second normally through a 1 m^2 area. So when discussing the apparent brightness of stars, we should in fact be quoting flux values.

However, when was the last time you heard a friend at your local astronomy club quote flux values in watts per square meter when, for example, talking about a newly discovered nova? Of course, when discussing the brightness of anything in the sky, he or she would use the word *magnitude*. The magnitude scale is something that all astronomers soon become familiar with; it consists of a sequence of simple numbers that are used to represent the brightness of objects in the sky. This makes it easy to remember the magnitudes of objects such as the planets, well-known stars, and deep sky objects; and it certainly beats having to deal with flux values. Even so, the magnitude of any object must be related in some way to the amount of flux that we receive from it. What is really amazing is that the magnitude scale enables the brightness of everything in the sky from the Sun to the faintest, farthest galaxy to be covered by a very small range of numbers. This feature alone makes it worthwhile getting to know the magnitude system in more detail.

From Flux to Flux Ratio to Magnitude

Professional astronomers will sometimes go to a lot of trouble to determine the actual amount of the total radiative flux, which we receive from a star. For reasons that we'll elaborate on in the next chapter, it is very difficult to do this, though it has been done for some stars; however, it turns out to be very much easier to *compare* the flux value of one star with that of another. This process is the basis of *astronomical photometry*, and it does not involve knowing the actual flux values themselves.

This might at first seem to make no sense; however, if you think about it, a visual variable star observer does his or her job by comparing the apparent brightness (in effect, comparing the flux values) of a variable star with that of one or more comparison stars. There is no knowledge of any actual flux values involved. To be sure these comparison stars have predetermined magnitudes, which enable the variable star observer to arrive at an estimate of the variable's magnitude, but it actually isn't even necessary to know the comparison stars' actual magnitudes in advance in order to be able to carry out the observation. Indeed, on variable star charts produced by the Variable Star Section of the British Astronomical Association, comparison stars are identified simply by letters to avoid any possibility of personal bias in making the observations.

So determining actual flux values of stars is hard, but comparing them with each other is relatively easy, and by "comparing flux values," we actually mean determining *flux ratios*. We are, in effect, saying that star "A" appears three times as bright as star "B" or that star "C" appears half as bright as star "D," and so on. It might at first seem possible that by choosing a particular star to serve as some sort of "standard star," one could define the "brightness" of any other star simply by the ratio of its flux value to that of the standard. However, the range of brightness for stars is enormous; the brightest stars appear many thousands of times brighter than the faintest ones, so while flux ratios are relatively easy to measure, because they span a very large range of values they are not practical for direct use as a measure of the brightness of a star. What we need, in fact, is a way to turn this large range of numbers into a small range of numbers.

The key to doing this lies in the well known party trivia question which says that if you take a standard chess or checkers board with 64 squares and place one coin on the first square, two coins on the second square, four on the third, and so on, how many coins would there be on the 64th square? The astounding answer turns out to be over 9 million trillion! Think about what we've done here, though; the numbers 1–64 cover a very modest range, yet the range of numbers covered by the quantity of coins on the squares is truly enormous. What's more, simply by knowing how far apart two squares are and knowing the rule that in this case we double the number of coins for each square moved up the board or halve the number for each square moved down, we can easily calculate the ratio of the number of coins on the two squares. Notice that we don't need to know how many coins there are on a given square. In this party piece example, think of the numbered squares of the chess board as representing "magnitudes" – a modest range of numbers. Think of the number of coins on a square as representing actual flux values – these numbers cover an enormous range. Finally, think of the "double the number of coins" rule as representing the flux ratios between stars of different magnitudes. This is basically how the stellar magnitude system itself works, but to get to its present form, it had to deal with its own history.

The ancient Greek astronomer Hipparchus classified the naked eye stars according to their brightness; the brightest stars were called stars of the first magnitude – this, in fact, amounted to about 20 or so stars. The next brightest group were called second magnitude stars, and so on, until the faintest stars visible to the naked eye were called stars of the sixth magnitude.

Hipparchus' idea to divide the naked-eye stars into six groups might be an interesting subject for historical discussion. Not long ago there was a talk at a local astronomy club, about the Greek writer Homer. The speaker said that according to Homer, Greek society at the time was divided into six social classes, which may have influenced Hipparchus to divide stellar brightness into six classes or magnitudes rather than, say, five or ten. Anyway, this could be a good research exercise for those keen historians among you. The fact is that Hipparchus' scheme stayed with us right through to the 19th century, by which time not only had vast numbers of telescopic (i.e., fainter than sixth magnitude) stars been discovered, but observational methods, both visual and instrumental, for comparing the brightness (in effect, determining flux ratios) of stars had been developed. It was time to bring the magnitude system into the modern age.

The English astronomer Norman Robert Pogson did this in the 1850s; he did it in a way which basically kept Hipparchus' division of naked-eye stars into six magnitude groups, and it also kept the tradition that a larger magnitude value corresponded to a fainter star.

The first thing to be clear about is that all of the stars that Hipparchus would have classed as being of sixth magnitude are not all of exactly the same brightness, but fall within a range of brightness that is itself embraced by Hipparchus' sixth magnitude group. Likewise, the first magnitude stars cover a range of brightnesses, but they still fall within the first magnitude group. By systematically comparing the brightness of stars, Pogson discovered that a first magnitude star was approximately 100 times brighter than a sixth magnitude star. He then went on to refine Hipparchus' magnitude groupings of stars into an actual number scale, *the magnitude scale*, by stating that a star *defined to be* of magnitude "1.0" is exactly 100 times brighter than a star defined to be of magnitude "6.0."

Here, then, is our first flux ratio: we receive 100 times as much flux from a star of magnitude (*mag.* for short) 1.0 than we do from a star of mag. 6.0. A further consequence of having a magnitude scale rather than a set of magnitude groups is that we can be much more accurate when specifying a star's brightness in terms of its magnitude; for example, rather than a "second magnitude star" we may have a star of mag. 2.34. What we need to do now is fit Hipparchus' second, third, fourth, and fifth magnitude groups into Pogson's magnitude scale, which, in keeping with the Hipparchus tradition, runs backward from brighter stars to fainter stars as the numbers on the scale increase.

What Pogson showed was that a *difference* of two numbers on the magnitude scale corresponds to a *ratio* of brightness or flux, and a magnitude difference of exactly 5.0 means a flux ratio of exactly 100. What about the flux ratio corresponding to a magnitude difference of 1.0? The first thing to say is that this flux ratio *must be the same* wherever we are on the magnitude scale; so a mag. 1.0 star and a mag. 2.0 star have the same flux ratio as, say, a mag. 4.0 star and a mag. 5.0 star, and so on; because we're looking for just the one number we shall call "x."

Let's imagine six stars that conveniently have magnitudes of 1.0, 2.0, etc., through 6.0, and list them in order of decreasing brightness together with the flux ratio "x" between each star and its next brighter and fainter neighbor:

Mag.	Flux ratio
1.0	x
2.0	x
3.0	x
4.0	x
5.0	x
6.0	x

What we see here is that it takes x stars of mag. 2.0 to produce as much flux as one star of mag. 1.0. In turn, it takes x stars of mag. 3.0 to produce as much flux as each of those mag. 2.0 stars. So it will take $x \times x$, or x^2 stars, of mag. 3.0 to match one star of mag. 1.0. And a magnitude difference of 2.0 corresponds to a flux ratio of x^2 whatever "x" is, and by following the same line of reasoning a magnitude difference of 5.0 means a flux ratio of x^5.

However, we know that this flux ratio is exactly 100 so x^5 equals 100; "x" then is equal to the fifth root of 100, i.e., that number which, when multiplied by itself five times, equals 100. The quick way to find "x" is to reach for your calculator and enter the number "100"; then press the "x^y" (or "y^x") key followed by the number 0.2. On pressing the "=" key you will see that "x" is a recurring decimal that, to three decimal places, equals 2.512. This number lies at the very heart of the magnitude system, and you should try not to forget it, because the number 2.512 raised to any power, which is itself a magnitude difference, results in the flux ratio corresponding to that magnitude difference.

With the magic number 2.512 (or, if you want greater accuracy, use the number your calculator actually gives for the fifth root of 100), we can easily work out flux ratios for any magnitude difference. Just enter the number 2.512 followed by the "x^y" key followed by the magnitude difference and finally followed by the "=" key to get the answer. For example, a magnitude difference of 3.0 means a flux ratio of $2.512^{3.0}$, which equals 15.85. You can even work out flux ratios for magnitude differences that are not whole numbers; for example, a magnitude difference of 2.83 gives us a flux ratio of $2.512^{2.83}$, which equals 13.55. This is just the kind of activity to take your mind off things, while sitting in the dentist's waiting room. It can also be useful for working out the brightness ratio of a variable star between its maximum and minimum magnitudes. For example, the long period variable star Chi or χ Cygni varies in magnitude from around 5 to around 15, a difference of 10 magnitudes. The flux ratio between the star's maximum and minimum is equal to 2.512^{10}, or about 10,000. So this is how a large flux ratio can come from a fairly modest magnitude difference.

Suppose, though, that you have actually measured the flux ratio between two stars; this is basically what CCD photometry software does. How do you then work out the corresponding magnitude difference? If one of the stars is a comparison star whose magnitude is known and the other a variable star, then this would enable you to calculate the magnitude of the variable. Sorting out this problem will also explain why the magnitude scale is often (sometimes perhaps rather too casually) referred to as a logarithmic scale.

How to Calculate Magnitudes

What we have so far is that for any two stars whose magnitudes we'll call "m1" and "m2" the value of 2.512^{m2-m1} equals the flux ratio for the two stars. The inverse of the problem is, given a flux ratio, what is the corresponding magnitude difference? For our two stars we'll let "m1" be the brighter star so that "m2 – m1" is a positive number. The flux ratio, we can call "f1/f2"; we don't know the actual values of "f1" and "f2," but we are assigning "f1" to the brighter of the two stars so that "f1/f2" will be a number greater than one. The basic magnitude formula now tells us that

$$2.512^{m2-m1} = f1/f2 \tag{2}$$

So, if we're given "f1/f2," how do we find "m2 – m1?" The simplest way is to take the logarithm of both sides of this equation. But before we do that, try the following simple example on your calculator. Enter the number 5 and then press the "LOG" key to get the logarithm of 5. Now multiply this number by 2; this should give you a number such as 1.39794, which you should then make a note of or store in your calculator's memory. After pressing the clear button, enter the number 25, which of course equals 5^2; again, press the "LOG" key and lo and behold you will see that this is the same number that you got before. So what we have shown is that $2 \times$ Log(5) is the same a $Log(5^2)$; this is a general mathematical rule, so we can also say that $Log(2.512^{m2-m1})$ is the same as $(m2-m1) \times Log(2.512)$. This means that if we take the logarithm of both sides of Equation (2) we have

$$(m2 - m1) \times \log 2.512 = \log(f1/f2) \tag{3}$$

The Log of 2.512 (or more precisely the fifth root of 100) is equal to 0.4, so

$$0.4 \times (m2 - m1) = \log(f1/f2) \tag{4}$$

and the proverbial slight rearrangement gives

$$m2 - m1 = 2.5 \times \log(f1/f2) \tag{5}$$

In other words, the magnitude difference between two stars is directly related to the *logarithm* of the ratio of the flux values; this is why the magnitude scale is termed a "logarithmic scale." Finally, if "m1" is a comparison star whose magnitude is known and "$m2$" is a variable whose magnitude we need to know, then we have

$$m2 = m1 + 2.5 \times \log(f1/f2) \tag{6}$$

One point regarding Pogson's excellent magnitude scale remains to be cleared up. We said above that Pogson *defined* a star of mag. 1.0 as being exactly 100 times as bright as a star of mag. 6.0. Among all the first magnitude stars, which one if any should we choose to be of exactly mag. 1.0? Or maybe we should pick a star to have by definition a magnitude of exactly 0.0; either way we need to "fix" the magnitude scale so that it has a "zero point."

Fixing the Magnitude Scale

There is an interesting and yet very important feature of the magnitude scale whose significance will become clearer in the next chapter. Because the magnitude scale is really all to do with flux ratios, which themselves are related to differences on the scale rather than actual flux values, it doesn't in principle matter where, within the vast population of stars, you actually put the zero point. For example, a seemingly logical choice might be to define the brightest star in the night sky, Sirius, as having a magnitude of exactly 0.0. This would simply mean that Sirius was 2.512 times brighter than a star defined to be of mag. 1.0 and 100 times brighter than a star defined to be of mag. 5.0 on this particular version of the magnitude scale.

The effect of choosing Sirius as the zero point would, however, make the fainter stars of Hipparchus' first magnitude group end up as second magnitude stars and so on, until the very faintest naked-eye stars would now come in at around mag. 7. Pogson himself in the interest of maintaining the Hipparchus tradition chose Polaris, a star classed by Hipparchus as being of the second magnitude to be of exactly mag. 2.0, until it was discovered that Polaris is slightly variable. The result was that Vega became the officially adopted zero point of the magnitude scale and was defined to have a magnitude of exactly 0.0 (for our present purpose this is fine, but as we shall see in the next chapter, defining a zero point for a magnitude scale is actually a bit more involved).

So Vega is now defined to be 2.512 times as bright as a mag. 1.0 star, and it also means that a star of mag. −1.0 is 2.512 times as bright as Vega; on this scale, Sirius now has a magnitude of −1.47. The Sun and the Moon, the brighter planets, and the very brightest stars have negative magnitudes on this scale, and in keeping with Hipparchus, the faintest naked-eye stars come in at around mag. 6.0. With a fixed zero point, it's then possible to establish a set or several sets of "secondary standard stars" that cover a whole range of magnitudes and which can be used to determine magnitudes for still more stars. The most well-known set of such standard stars is the North Polar Sequence, a set of about 100 stars that are distributed around the north celestial pole.

The final thing to say here is that the magnitude scale is of course based on how bright stars appear in the sky, either to the naked eye or as seen visually through a telescope; it is thus more correctly referred to as the *visual magnitude scale*. The magnitude of a star or any other object on this scale is also referred to as the object's *apparent magnitude* and such magnitudes are always denoted with a lower case "*m*." The apparent magnitude of a star depends on the amount of radiative flux that we receive from it, and this in turn depends on both its luminosity and its distance. By contrast a "magnitude" that did not depend on distance would depend only on a star's actual luminosity.

Absolute Magnitude – A Measure of Stellar Luminosity

If all stars could be lined up at the same distance from Earth, then their apparent magnitudes would depend only on their luminosities. In 1922 the International Astronomical Union decreed that such a distance should be 10 pc, or 32.6 light years, and the apparent magnitude a star would have if it were this distance from Earth is called the *absolute magnitude*, which is always denoted by a capital "M." The absolute magnitude scale works in exactly the same way as the apparent magnitude scale. As with the apparent magnitude scale, a magnitude difference of 1.0 on the absolute magnitude scale also corresponds to a brightness or flux ratio of 2.512.

In our thought experiment with the Dyson sphere, we saw that the flux received by each square meter of the inner surface of the sphere was simply equal to the star's luminosity "L" divided by the surface area of the sphere, or $L/4\pi r^2$, "r" being the radius of the sphere. If the distance from the star to Earth is "d," then clearly the flux that we receive is equal to $L/4\pi d^2$; we shall call this flux "f" and assume that "d" is measured in parsecs. If this star were now "moved" to a distance of 10 pc then the flux we receive would be $L/(4\pi \times 10^2)$ or $L/(4\pi \times 100)$; we shall call this flux "F." So the ratio of these two flux values is given by

$$\frac{F}{f} = \frac{\frac{L}{4\pi \times 100}}{\frac{L}{4\pi \times d^2}} = \frac{d^2}{100} \tag{7}$$

From the basic magnitude formula (Equation 5), the number 2.5 multiplied by the logarithm of this ratio is equal to $m - M$, i.e., the difference between the apparent and absolute magnitudes for the star.

Now here's another little example to try on your calculator. Enter the number "2" and press the "LOG" key to give log 2 and store the number or make a note of it. Now determine log 5 in the same way and subtract log 2 from it; you should get something like 0.39794. Finally, determine log 2.5 – log 5/2 – and you will get the same answer; so log 5/2 is the same as log 5 – log 2, and again this is a general rule. So now we can say that log $d^2/100$ is equal to log d^2 – log 100; log 100 equals 2, and remember from our previous calculator example that log d^2 is equal to $2 \times \log d$. So

$$m - M = 2.5 \times (2 \log d - 2) \tag{8}$$

and

$$m - M = 5 \log d - 5 \tag{9}$$

This simple but very important formula is called the *distance modulus formula* and the difference between the apparent and absolute magnitudes of a star is called the star's *distance modulus*; remember that the distance to the star "d" must be in parsecs.

Distance moduli for stars divide very simply into two groups. A star whose distance is more than 10 pc will have an apparent magnitude fainter than its absolute magnitude, and so $m - M$ will be a positive number. By contrast, the apparent magnitude of a star closer than 10 pc will be brighter than its absolute magnitude, and $m - M$ will be negative; a star at a distance of exactly 10 pc will, of course, have a distance modulus of zero. Finally, here are a few simple distance modulus examples.

First take Sirius, the brightest night time star with an apparent magnitude of $m = -1.47$. Sirius' distance of 8.7 light years converts to 2.67 pc. So plugging these numbers into Equation (9) gives us

$$- 1.47 - M = 5 \log 2.67 - 5$$

Rearranging gives us Sirius' absolute magnitude as

$$M = - 1.47 - \log 2.67 + 5$$

This gives Sirius an absolute magnitude of +1.4; so at 10 pc it would be just about classed as a first magnitude star.

Rigel (β Orionis) has an apparent magnitude of 0.08 and an estimated absolute magnitude of –7.0, so its distance modulus is equal to 0.08 – (–7.0), which is 7.8. So using the distance modulus formula, Equation (9), 7.8 = 5 log d – 5; and so, log d = 12.8/5 = 2.56 where d is Rigel's distance in parsecs. The antilog of 2.56 (on your calculator; enter 2.56 then press the "INV" button followed by the "LOG" button) gives Rigel's distance "d" as about 363 pc, which when multiplied by 3.26 gives an approximate distance to Rigel of 1,200 light years.

Finally the Sun has an apparent magnitude of –26.74 to two places of decimals, and its distance is 1 AU; 1 AU is equal to 4.8544×10^{-6} pc. So Equation (9) gives us

$$- 26.74 - M = 5 \log (4.8544 \times 10^{-6}) - 5$$

and so the Sun's absolute magnitude is equal to

$$M = -26.74 - 5\log\left(4.8544 \times 10^{-6}\right) + 5$$

Remember, to enter a number such as 4.8544×10^{-6} first enter 4.8544 and then press the key marked "EXP." Now press the key marked "+/–" followed by the number 6. On pressing the log key you should get –5.314 to three decimal places. Multiply this by 5 to get –26.57, and we now have

$$M = -26.74 - (-26.57) + 5$$

which equals +4.83.

So at a distance of 10 pc, our Sun would be a dim star of mag. 4.83.

The important thing about doing calculations like this is to take your time and be careful to get the combination of plus and minus signs right. The other thing is to use your common sense; if a star is further away than 10 pc then its absolute magnitude must be brighter than its apparent magnitude, and conversely if it's closer than 10 pc, its absolute magnitude will be fainter than its apparent magnitude.

The magnitude system is a brilliantly succinct and simple way to represent brightness or flux ratios between stars; however, since Pogson's time, things have moved on yet again, and as we shall see in the next chapter, the magnitude system itself has evolved into an even more refined tool for investigating starlight.

Key Points

- The overall flow of radiation through a region of space is called a radiation field.

- The net rate of flow of energy through $1\,m^2$ in the radiation field is called the flux, which is measured in watts per square meter.

- If the flux is the same in all directions, then the radiation field is said to be isotropic.

- If the total radiant energy per unit volume is the same everywhere at all times, the radiation field is said to be homogeneous.

- In practice, radiation fields are non-homogeneous and anisotropic.

- The intensity of a radiation field is the rate of flow of energy through $1\,m^2$, which is confined to a very narrow range of directions.

- Intensity does not change with distance from the source of the radiation whereas flux does.

- The total power output of a star, measured in watts, is called its luminosity.

- The apparent brightness of a star as seen from Earth is determined by how much flux we receive from it.

- The apparent magnitude of a star is a measure of how much flux we receive from it.

- The absolute magnitude of a star is the apparent magnitude it would have; that is, it is a measure of how much flux we would receive if the star were at a distance of 10 pc.

- The distance modulus of a star is equal to the difference between its apparent magnitude "m" and its absolute magnitude "M."

A Multitude of Magnitudes
for the Colors of Starlight

There can be no doubt that one of the most significant ongoing contributions that amateur astronomers make to the science of astronomy is the regular observation of variable stars. Most amateur observers make visual observations that basically involve "bracketing" the variable between a slightly brighter and a slightly fainter comparison star in order to arrive at an estimate of the variable's magnitude. A very large number of the stars observed by amateurs are Mira-type long-period variables and semi-regular or irregular pulsating stars. Most of these stars are red, and one of the first things that a visual observer learns is to make their estimates using short glimpses, rather than long "stares," because of the way in which the human eye responds to red light. Red stars appear to grow brighter the longer one stares at them – a phenomenon called the "Purkinje effect." Even so, it has been said that trying to estimate the magnitude of a red star by having to compare it with white comparison stars (as is often the case) is like trying to compare the strength of one cup of tea with the temperature of another.

For the increasing number of variable star observers who use a CCD camera to make instrumental magnitude estimates, the situation becomes even more acute. Surely it has been called into question as to whether one should always use special photometric filters when making observations. After all, if you're trying to get an early image of an outbursting nova that is still very faint, you don't want to cut down your camera's sensitivity by sticking a filter in front of it. On the other hand,

K. Robinson, *Starlight*, Patrick Moore's Practical Astronomy Series,
DOI 10.1007/978-1-4419-0708-0_5, © Springer Science+Business Media, LLC 2009

it's expected that novae are likely to change color during the course of their outburst; the result is that unfiltered observations become of limited value, and more generally you'll find that CCD variable star observers are strongly recommended to use special color filters when making their observations. All of this is telling us that this magnitude business needs even further investigation, and that's what we'll do here. This will in fact tell us a great deal about stars themselves and not just whether or which color filter to use.

Blackbody Stars

As we saw in chapter *From Light to Starlight* stellar spectra resemble those of blackbodies. This means, of course, that like blackbodies, hotter stars will emit more radiation at shorter wavelengths and will thus appear bluish, whereas cooler stars will emit most of their radiation at longer wavelengths and will appear red. There are differences between the spectra of stars and blackbodies, but the overall shape of the continuum part of a star's spectrum is that of a curve, which rises relatively steeply on the short-wavelength side to a maximum value and then falls more slowly on the long-wavelength side. The differences are there, first because of the presence of absorption lines and maybe also emission lines in the spectra of stars. There are also differences within the continuum itself, which are essentially due to "things" going on in the outer layers of stars and maybe also in their immediate environment.

As you might expect, working out the true cause of these extra features present in stellar spectra is one of the most important and most fascinating areas of stellar astronomy (it can also prove to be very complicated). As always, though, it's best to start by doing things in a simple way, and with stars this means assuming that they radiate just like perfect blackbodies. We can call such approximated stars "blackbody stars."

In chapter *From Light to Starlight* we also saw that the amount of energy emitted every second (in other words the total flux "F") from each square meter of the surface of any blackbody, and that includes the surface of our "blackbody star," is given very simply by the Stefan–Boltzmann equation, i.e.,

$$F = 5.670 \times 10^{-8} \times T^4 \,\mathrm{W/m^2} \qquad (1)$$

T is the temperature in degrees Kelvin, and the number 5.670×10^{-8} is known as Stefan's constant; so if we could measure the total flux from a square meter of a star's surface, then we could determine its temperature. Another, perhaps even simpler way to "take a star's temperature" is to note from the star's spectrum the wavelength λ_{max} at which maximum flux is being emitted. We can then use Wien's law, which says

$$\lambda_{max} = 2.8973 \times 10^7 / T \qquad (2)$$

A simple rearrangement will give

$$T = 2.8973 \times 10^7 / \lambda_{max} \qquad (3)$$

Once again the temperature T is in Kelvin and the wavelength λ_{max} is in angstroms.

So straightaway by adopting this blackbody approximation we appear to have not just one but two simple ways to determine or at least make a reasonable estimate of a star's temperature. This is an important number, so let's give it a go and see if Equations (1 and 3) give us the right kinds of answers.

If we try Wien's law first, λ_{max} for the Sun is about 5,000 Å, and dividing 2.8973×10^7 by 5,000 gives 5794.6 K, which is wonderfully close to the figure of 5,800 K that most of us will have come across at some stage when reading about the Sun's surface temperature. Stellar temperatures determined in this way are perhaps not surprisingly referred to as "*Wien temperatures.*"

Before becoming overconfident, though, let's try using the Stefan–Boltzmann law to measure the Sun's temperature. This is a bit more involved, because we need to know how much flux is radiated from 1 m^2 of the Sun's surface. The most straightforward way to do this is to simply divide the total flux from the entire Sun (i.e., the Sun's luminosity) by the Sun's surface area. The Sun's luminosity as stated in chapter *Space – The Great Radiation Field* is equal to 3.827×10^{26} W, and its surface area ($4\pi \times R_\odot{}^2$; where "$R_\odot$" is the Sun's radius) is equal to 6.087×10^{18} m^2. So dividing the luminosity by the surface area gives 6.287×10^7 W/m^2. The Stefan–Boltzmann equation tells us that this number divided by Stefan's constant (using your calculator to do this you should get 1.109×10^{15}) is equal to the Sun's surface temperature raised to the power four.

With the number 1.109×10^{15} entered on your calculator simply press the square root key ("$\sqrt{}$" or possibly "INV" followed by "x^2") twice. This should give you about 5,770 for the temperature in Kelvin, again very close to that figure of 5,800 K. The temperature of a star calculated using the Stefan–Boltzmann equation is called the *effective temperature*, and it is usually written as T_{eff}. So according to our calculation the effective temperature of the Sun is about 5,770 K.

We seem to be doing remarkably well, considering that we're using fairly simple equations to calculate something as important as the temperature of a star. Let's see if it works on a distant star, such as for example Vega; this is clearly more difficult, because we need to know the distance to the star in order to determine its luminosity and we also need to know its radius (and incidentally assume that it is spherical) in order to be able to calculate its surface area and so work out the flux from 1 m^2 of its surface.

The luminosities and radii for a reasonable number of stars including Vega are actually known, though we won't go into the details here of how they have been determined; we simply want to test how good our blackbody approximation is for stars. The luminosity of Vega is equal to about $37L_\odot$, or about 1.416×10^{28} W. Vega is not spherical, but its average radius of $2.52R_\odot$ gives us an approximate surface area of 3.866×10^{19} m^2. Again, using the Stefan–Boltzmann equation and following exactly the same procedure as we did for the Sun gives us a surface temperature of 8,965 K; this is about 1,000 degrees below Vega's accepted effective temperature of around 10,000 K, but it's not bad considering we used one or two approximate values in our calculation.

The wavelength of maximum flux for Vega (and indeed stars like Vega) is around 4,200 Å, and if we apply Wien's law to this wavelength, then we get a temperature of only around 7,000 K, which is clearly way off the mark. If we think about it, though, this result should not be entirely surprising; we know that stars in reality do not radiate as perfect blackbodies, so we cannot assume that the wavelength of maximum flux for a real star like Vega is the same as it would be if Vega were a blackbody (in chapter *First Look Inside a Star – The Atmosphere* we'll see why using the Wien's law gives such a poor result for stars like Vega).

The calculation that used the Stefan–Boltzmann equation to work out Vega's effective temperature seemed to be more successful than using Wien's law, and indeed when astronomers refer to the "temperature" of a star it is usually the effective temperature that they are talking about. As we saw above, though, calculating the effective temperature of a star, or a fairly fundamental property of any star, requires prior knowledge of some other equally fundamental properties of the star – in particular its luminosity and its radius. What we need is a reliable way to calculate effective temperatures for stars using real astronomical observations, rather than fundamental physical properties, which are usually hard to come by, and ideally we want those astronomical observations to be relatively simple to carry out.

From Theories of Physics to Astronomical Observations

Up to now we have been perhaps a bit casual in our use of the word "flux"; the calculations that we have performed involved using the "total flux," i.e., the entire quantity of flux emitted across all wavelengths of the electromagnetic spectrum. Although, as mentioned in the previous chapter, this has been determined for some stars, it is exceedingly difficult to do.

In the world of real astronomy, where life seldom runs like a physics textbook, the "flux" from a star comes in the form of the star's apparent magnitude, which as we saw in chapter *Space – The Great Radiation Field* is determined by comparing the star to one or more "standard stars." Furthermore, necessity forces astronomers to use magnitudes that in fact do not represent flux values spanning the entire electromagnetic spectrum but only a limited part of it, that part received and measured by our telescopes and detectors. After all, even Pogson's formalized scale of magnitudes involved only that part of the flux from a star that could be detected by the human eye.

In fact, in the mid-19th century this *visual magnitude scale* was all that was needed when dealing with the brightness of stars. However, things began to change when photography was used to make images of stars, because it was soon realized that the sensitivity of a photographic plate was different from that of the human eye. To be more specific, a photographic emulsion worked well in the blue part of the spectrum and even in the near ultraviolet, but its sensitivity fell off dramatically in the red; older black and white photographs of Orion, for example, show a miserably faint Betelgeuse compared to most of the other predominantly blue stars. By contrast the maximum sensitivity of the eye is centered at around 5,000 Å in the green part of the spectrum and falls away quite sharply toward both the blue and the red. The bottom line is that the brightness, or the magnitude of a star as determined by the eye, is different to that determined by a photographic emulsion.

It thus became clear that astronomers had to be rather more concise when referring to the apparent magnitude of a star than they had hitherto been. Previously, there had only been eyes; then there were two detectors – eyes and photographs – which meant having two magnitude scales and hence two possible magnitudes for each and every star. The diameter of a star's image on a photograph could be measured and compared with those of "standard stars" (these were established in the same way as those on the visual magnitude scale) to give the star's *apparent photographic magnitude*, usually designated m_{pg}. As with the visual magnitude scale the standard stars themselves serve as reference points on the *photographic magnitude scale*, which was established by the American astronomer and co-founder of the American Association of Variable Star Observers, Edward Charles Pickering. On this scale white stars such as Vega and Sirius were found to have visual and photographic magnitudes that are more or less equal (so, for example, Vega would have both visual and photographic magnitudes equal to 0.0).

For stars that are not white, the two magnitudes will be different. What this means, for example, is that a blue star such as Rigel will show up relatively brighter on a photograph and so its photographic magnitude will be a smaller number than its visual magnitude. By contrast the relatively faint photographic magnitude of a red star such as Betelgeuse will be a bigger number than that of its visual magnitude. So straightaway we can see here that the difference between a star's two magnitudes serves as a measure of the star's color, and, if we think in terms of blackbody stars, this also means temperature.

The subsequent inclusion of light sensitive dyes into photographic emulsions extended their sensitivity into longer wavelengths and enabled them to more closely mimic the spectral response of the eye, provided they were used in conjunction with a yellow filter (which filters out the excess shorter wavelength radiation affecting a photographic emulsion but not the eye). Magnitudes measured in this way more closely resemble visual magnitudes and were called *photo-visual magnitudes* (denoted m_{pv}). Thus it was possible to have two photographically determined magnitudes for a star, one that essentially measured light in the green part of the spectrum and the other in the blue. The difference between the photographic and photo-visual magnitudes (i.e., $m_{pg} - m_{pv}$) was called *the color index* of the star or sometimes the *international color index*.

As before, a blue star would have a brighter, or lower numerical value for m_{pg} than for m_{pv}, and so its color index would be a negative number, whereas for red stars (which would be relatively brighter at longer wavelengths) the color index would be a positive number. Finally, somewhere in between there would be stars such as Vega with equal photographic and photo-visual magnitudes whose color index equaled zero. The term "color index" now has a more modern meaning, which we're very shortly coming to; however, you may still occasionally come across this old form of the color index in the literature.

In its time the great thing about measuring photographic and photo-visual magnitudes was that a large number of stars could be measured on one single image, and this method continued right up until the latter part of the 20th century. However, even during the opening years of the 20th century, the American astronomer Joel Stebbins was pioneering the use of what were called "photosensitive cells," which used the (then just freshly explained by Albert Einstein) photoelectric effect to turn starlight into a measurable electric current.

Over half a century, photosensitive cells gave way to photomultipliers, which when attached to telescopes became the standard instruments for measuring the magnitudes of individual stars. *Photoelectric photometry*, as the technique is known, proved to be capable of determining magnitudes to a far greater degree of accuracy than either photographic or visual methods, but it meant that astronomers had to think yet again about the meaning of the word "magnitude."

Star Partners – Color and Magnitude

The old idea of the international color index, which in a sense came about by chance, nonetheless proved to be extremely useful, because a simple, easily determined number gives a measure of the color of a star. More specifically, it is a measure of how much radiation a star is emitting at the blue end of the visible spectrum, compared to that in the middle (i.e., yellow/green region), and, as we know, more emitted radiation at blue wavelengths means that the star is hotter. This means that the color index itself must be a measure (yet another one) of the star's temperature. This was recognized as being very significant, because if a relation could be found between the color index of a sample of stars and their effective temperatures, this relation would then provide a very simple means of determining the effective temperatures of other stars, without having to know their luminosities or their surface areas. It would only be necessary to measure a star's magnitude at two different wavelengths, and what's more, the use of photomultipliers meant that magnitudes could be measured quickly, easily, and very accurately.

The much greater sensitivity of photomultipliers meant that astronomers could be choosier when thinking about the color index of a star. By using specially manufactured color filters, the choice was in fact not only one of which regions of the spectrum to use for defining star magnitudes but also how wide in terms of what range of wavelengths covered those regions should be. The result was a plethora of so-called *photometric systems*. Those which employed regions of the spectrum whose widths were less than 90 Å were perhaps not surprisingly called *narrow-band systems*. In turn, *intermediate-band systems* involved regions of widths 90–300 Å, and finally *wide-band systems* used regions of widths greater than 300 Å. Clearly a narrow-band system will effectively analyze starlight in more detail than an intermediate system or a wide-band system, but the downside is that more filters and of course more actual observations are needed to do the job.

Most photometric systems are used for specialist purposes; for example, narrow-band systems may be used to investigate individual absorption or emission lines in the spectra of stars. The most widely (and for the amateur, more or less exclusively) used photometric system is a wide-band system established by American astronomers H. L. Johnson and W. W. Morgan in the early 1950s. This system initially focused on three regions of the spectrum, each of width around 1,000 Å, which were "isolated" by using standardized photometric filters. Two of these filters, the "B" for blue (centered at around 4,400 Å) and the "V" for "visual" (centered at 5,500 Å) effectively provided photoelectric photometry versions of the old photographic and photo-visual magnitudes, respectively. The third, "U," for ultraviolet filter (at 3,600 Å), took advantage of the photomultiplier's sensitivity in the violet and near ultraviolet part of the spectrum. A series of plots showing how much radiation is transmitted by each filter as a function of wavelength (maximum transmission is given the value "1" in each case) is shown in Fig. 1, each plot being called the *response function* of the filter. The actual location of the "U" and "B" regions was also carefully chosen for another reason, which we'll come to later.

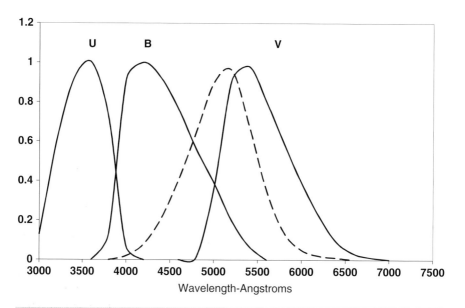

Figure 1. The filter "response functions" for the Johnson and Morgan "UBV" photometric system. The *dashed curve* gives the approximate response function of the human eye.

So the Johnson and Morgan photometric system gives us three different magnitude scales and three magnitudes – m_U, m_B, and m_V – for every star, and note also that as with the traditional visual magnitude scale, we can also define absolute U, B, and V magnitudes (written M_V, etc.) as the apparent U, B, or V magnitude that a star would have at a distance of 10 pc. This "UBV" system enables two separate color indices to be defined for stars; $m_U - m_B$ (often just written as U – B) simply involves taking the difference between the magnitude of a star as measured through the U and the B filters. Likewise, B – V involves doing the same thing with the B and the V magnitudes.

Notice here that the convention is to take the magnitude at the shorter wavelength and subtract from it the magnitude at the next longer wavelength; so, for example, a "U – V" color index is not generally used. As with the older international color index a negative value for B – V implies a relatively hot star with a brighter B magnitude, and furthermore the very hottest stars will also have negative values for U – B. Each of these magnitude scales works in exactly the same way as Pogson's original visual magnitude scale and as with the visual magnitude scale. The "U," the "B," and the "V" magnitude scales all have to have zero points or standard stars against which other stars can be compared.

Two possible scenarios suggest themselves here. The first would be to have a different standard star for each of the magnitude scales, and this would likely be a star whose peak emission wavelength corresponded with that of the magnitude scale itself. So, for example, we could choose Vega as the standard star for the B magnitude scale and Capella for the V magnitude scale. Let's see how this would work when it comes to dealing with color indices.

Vega would have mag. 0.0 on the B scale, but it would be fainter than Capella on the V scale (i.e., its V mag. would be a positive number, because being a hotter whiter star it emits less flux in the green part of the spectrum). So B – V for Vega and stars like it would be negative numbers. By contrast, Capella would have mag. 0.0 on the V scale but would be fainter than Vega on the B scale, and B – V for these kinds of stars would be positive numbers.

On the face of it, this seems okay, because it means that hotter stars will have negative values for B – V whereas cooler stars will have positive values. It also means that somewhere in between stars such as Vega and those like Capella there ought to be stars whose B – V value equals 0.0. Again, this might seem acceptable, but the crunch probably comes as a result of the fact that in recent years the Johnson and Morgan photometric system has been extended to include an "R" for red, an "I" for infrared, and also additional magnitude scales extending further into the infrared part of the e-m spectrum. This would mean finding additional standard stars, and with red stars we run into trouble, because most of them are variable; in particular two obvious candidates, Betelgeuse and Antares, are both variable. Even if this problem were overcome, it would still mean that there would be, in effect, a whole series of different star types for which the various color indices (B – V, V – R, etc.) are equal to zero, and clearly this would make it very difficult to try and relate color index to effective temperature. It's perhaps not surprising then that this method of defining color indices was not used.

The second scenario, the one that is used, takes just one star, Vega, to be the standard star for all of the photometric magnitude scales, so that on the U, B, V, R, I, etc., scales the magnitude of Vega is *defined to be* exactly 0.0. This means first of all, of course, that each color index (U – B, B – V, etc.) for Vega is equal to zero, and this also applies to all stars such as Vega, for example, those stars that have the same distribution of flux across the e-m spectrum as Vega itself. It also means that all color indices are defined with respect to this one star, and in consequence the value of any color index for a given star can tell us whether the star is hotter or cooler than Vega.

A star such as Capella, which is cooler than Vega, will as expected have a B – V value, which is positive, and what's more an even cooler star such as Betelgeuse will have an even larger positive value for B – V. By contrast, a star such as Rigel, which is hotter than Vega, will have a negative value for B – V, and the hotter a star is, the more negative the B – V color index will be.

Of course, what we have done here is to adjust each of the Johnson and Morgan magnitude scales to make Vega the zero point for that scale, but it's actually okay to do this, because the whole magnitude system "runs" on *comparing* flux values, rather than measuring *actual* flux values. So on each of the magnitude scales we are simply comparing any star we care to investigate with Vega.

Two final notes: the result of defining Vega to have mag. 0.0 in all the regions of the Johnson and Morgan photometric system resulted in Vega being "shifted" away from the zero point of the Pogson's visual magnitude scale, where it now comes in at mag. +0.03. Secondly, in the 1980s, observations by the InfraRed Astronomical Satellite showed that Vega had an "infrared excess" – in other words, it was brighter in a part of the infrared spectrum than a star of its kind should be. This, it was realized, was due to dust surrounding Vega, which is warmed by Vega itself and in consequence produces its own thermal radiation, which effectively contaminates Vega's spectrum. There is also the fact that in order to warm the dust some of Vega's radiation (at shorter wavelengths) is absorbed, thus further altering Vega's spectrum albeit maybe ever so slightly. Regarding where this leaves Vega as the "zero star" for all magnitude scales, the jury still seems to be out on this one.

Color and Blackbody Temperature

The effective temperature of a star is the temperature that the star would have if it were radiating as a perfect blackbody; more precisely, effective temperatures are calculated using the Stefan–Boltzmann law. Though stars are not perfect blackbodies, it would clearly be a good idea to investigate the B − V color indices for actual blackbodies, which of course would include blackbody stars of different temperatures. Because the theory of blackbody radiation is well understood, it's possible to calculate how much flux is radiated over the wavelength regions covered by, for example, the Johnson and Morgan B and V bands.

A thoroughly rigorous calculation is not easy, however, due in part to the shape of the response functions of the filters, but a good approximation is obtained by calculating just the flux emitted at the peak transmission wavelength for each filter. Even so, the calculations that make use of the Planck blackbody radiation formula mentioned in chapter *From Light to Starlight* are still pretty involved, so we won't go into the details here. Indeed, we don't need to, because what really matters is the ratio of these flux values rather than the values themselves, because as we saw in chapter *Space − The Great Radiation Field* and in particular in Equation (5), flux ratios convert to magnitude differences, and here these magnitude differences will give us the B − V color indices for blackbodies.

Table 1 lists the calculated B − V indices for a range of blackbody temperatures, and straightaway we can see that the B − V index for a blackbody temperature of 10,000 K (the effective temperature of Vega) is not equal to 0.0. Remember, though, that the magnitude scales for real stars have been adjusted so that the various color indices for stars such as Vega all equal 0.0, and we can do exactly the same thing here. For a blackbody temperature of 10,000 K (*and only* 10,000 K) we can define all of the Johnson and Morgan color indices to be equal to 0.0. In practice, all that we have to do is add the same small number to each of the B − V values for the various temperatures in order to ensure that B − V for 10,000 K equals 0.0. A similar small correction can be applied to the other color indices in the Johnson and Morgan system, to ensure that their values for 10,000 K also equal 0.0. What we've effectively done is to create a "true blackbody Vega" having a temperature of 10,000 K against which all blackbody stars could be compared.

Finally, if we plot the temperatures in Table 1 against the corresponding adjusted B − V indices, as shown in Fig. 2a, we see that we have a smooth simple curve that has no bumps or wiggles, and this means that for perfect blackbodies the value of B − V varies in a fairly predictable way

Table 1. Calculated B – V values for a range of blackbody temperatures. Column 3 gives the adjusted B – V values to ensure that B – V for a temperature of 10,000 K equals 0.0.

Blackbody temperature (K)	B – V	B – V adjusted
3,000	1.152693	1.614203
4,000	0.562975	1.024485
5,000	0.211457	0.672967
5,800	0.01975	0.48126
6,000	−0.01984	0.441669
8,000	−0.30088	0.160634
10,000	−0.46151	0.0
14,000	−0.63293	−0.17142
20,000	−0.74968	−0.28817

with temperature. In addition we have plotted in Fig. 2b the logarithm of the temperature against B – V. Remember that the Stefan–Boltzmann law tells us that the temperature of a blackbody is directly related to the flux emitted from the body's surface. In turn the log of the temperature will be related to the log of the flux which, as we saw in chapter *Space – The Great Radiation Field*, is effectively equivalent to a star's magnitude. Indeed, the shape of the curve in Fig. 2b is significant, as we shall see shortly.

The question now is, what would such a plot look like for real stars?

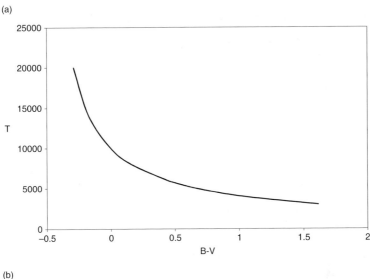

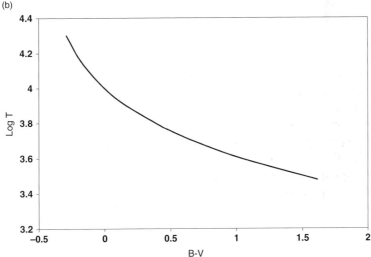

Figure 2. In **a** we have plotted temperature (in Kelvin) against B – V color index for perfect blackbodies. The *smooth shape* of this curve tells us that B – V varies in a very steady fashion with temperature. Even more revealing is the plot in **b** of Log (temperature) against B – V, which shows a striking similarity to the overall shape of the main sequence region of the Hertzsprung–Russell diagram (see below).

Measuring Effective Temperatures for Real Stars

It goes without saying that a tremendous amount of work has been done over the years (and it is indeed still going on) to determine effective temperatures for stars using the Stefan–Boltzmann law, so that they can be compared with the stars' observed color indices. As already outlined, three pretty fundamental pieces of information are needed about a star in order to determine its effective temperature. These are its distance, its radius (or diameter), and the total amount of flux that we receive from it across all wavelengths of the e-m spectrum.

The distance is probably the easiest piece of information to get at, and this has certainly been helped over recent years by the Hipparcos spacecraft. Determining a star's diameter is much more difficult; one method is to analyze the light curves of eclipsing binary stars, where in principle the duration of the minima in the light curve, which results from the passage of one of the stars in front of the other as seen from Earth, can give information on the size of the component stars. Although this method can certainly be successful, the main drawback comes from the fact that the components of an eclipsing binary system are often not what we might call "normal stars" (the meaning and significance of this will become clearer shortly) and so are probably not good candidates with which to try and establish some relation between effective temperature and color index.

A more direct method initially developed in the 1920s and nowadays used widely in conjunction with large telescopes employs an instrument called an interferometer, which is capable of resolving very small angular separations on the sky – small enough to be able to measure the apparent size of a star's visible disk. Provided the distance to the star is known, this disk size converts into an actual diameter for the star. (The method has even been used to discover, for example, that some stars, including Vega, are not spherical.)

This then leaves the business of determining the total amount of flux that we receive from the star, and this is probably the hardest thing of all to measure. It means first of all having an array of instruments that can measure flux in various wavelength regions. Some of these detectors (for example, those for X-rays and for some infrared and ultraviolet wavelengths) will need to be above Earth's atmosphere, which would otherwise absorb the incoming flux. Even so, there are some wavelengths that "go missing" because they are absorbed out there in interstellar space (flux values here have to be "filled in" using theoretical methods), and even those wavelengths that do reach ground-based detectors need to have measured flux values, corrected to account for absorption both in interstellar space and in Earth's atmosphere. However, total flux values have been determined for a big enough sample of stars to make a calibrated relation between effective temperature and B – V color index possible. The result of all this effort by what very likely amounts to a lot of unsung heroes is shown in Fig. 3; notice the similarity of this curve with that in Fig. 2b. This very reassuringly tells us that the continuous spectra of stars are indeed similar to those of blackbodies.

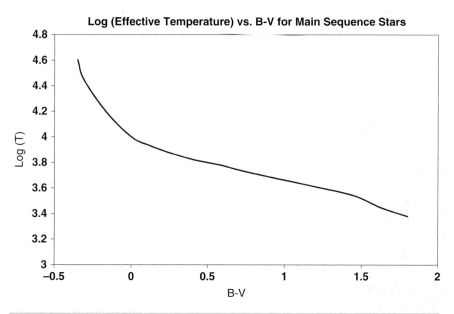

Figure 3. This plot of the log of the effective temperature against B – V for a sample of real stars shows a strong similarity to the curve in Fig. 2b and helps justify approximating stars to blackbodies.

On a final note, the B – V color index is a much better measure of the effective temperature for those stars that emit much of their radiation across the "B" and "V" regions. For cooler stars, for example, the shape of the continuum is much flatter over these regions, and a better indicator of temperature might be, for example, the V – R, the R – I, or even color indices further into the infrared.

Yet Another Magnitude

We are well familiar now with the idea that observed flux values for stars are represented by magnitudes on an appropriate magnitude scale. Figure 3 uses measured flux values for a sample of stars covering the entire e-m spectrum, and the magnitude scale that goes with these total flux values is called the *bolometric magnitude scale*. The *apparent bolometric magnitude* of a star on this scale is written as m_{bol}. As with Pogson's visual magnitude scale, we can also define the absolute bolometric magnitude, or M_{bol}, as the apparent bolometric magnitude that a star would have at a distance of 10 pc; in other words, M_{bol} is a measure of the total flux received from a star at a distance of 10 pc and so can be regarded as the true measure of a star's luminosity.

The apparent bolometric magnitude itself is not used very much at all, but it is interesting to note that unlike the case with all other magnitude scales, the apparent bolometric magnitude of Vega is not 0.0. The reason for this, which actually applies to most stars, is that bolometric magnitudes will always be brighter than either visual magnitudes or, for example, V magnitudes. This is simply because bolometric magnitudes include extra flux emitted at other wavelengths and so they are brighter. The result is that m_{bol} for Vega is equal to –0.3. The magnitude adjustment that results from taking into account the extra flux is called the *bolometric correction*, or "BC" – a term used a lot in the literature. Some authors define BC as being equal to V – m_{bol}, in which case BC is always a positive number, but more often m_{bol} – V defines BC, which is then always negative.

The significance of the bolometric correction is that the greater its value – either more positive or more negative – whichever convention you use, the more radiation the star will be emitting at wavelengths other than those in the visible part of the spectrum. This is the case with stars that are either very hot or relatively cool. Stars with middle-range temperatures emit most of their radiation in the visible part of the spectrum and so will have small BC values.

This immediately ties in the BC value for a star with its B – V color index, and the plot in Fig. 4 shows how the bolometric correction varies with B – V. The BC scale itself has by convention been adjusted so that stars with a B – V color index of about 0.42 (most of the radiation from these stars, which have an effective temperature of around 6,500 K, is in the visible part of the spectrum) have a bolometric correction of 0.0 (the bolometric correction for the Sun, by the way, is –0.07). This, in turn, of course will fix the zero point of the m_{bol} scale so that instead of having a "standard star," a star of m_{bol} equal to 0.0 is one that produces a total flux value of 2.48×10^{-8} W/m^2 at the top of Earth's atmosphere.

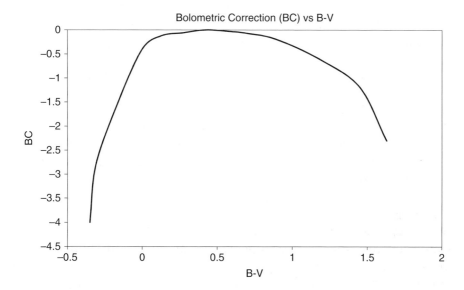

Figure 4. The bolometric correction "BC" vs. B − V for "normal" stars. The BC scale is by convention adjusted to equal 0.0 for a B − V value of around 0.48. Its increasing negative value on either side of this point results from both hotter and cooler stars emitting increasing amounts of radiation at wavelengths other than the visible region.

Photometric systems such as the Johnson and Morgan UBV system together with their associated color indices are clearly important and powerful tools, which are used very extensively these days by both professional and amateur astronomers. They effectively sample the light from stars and provide valuable and relatively easy to obtain information on the manner in which the radiation from a star is distributed across the various wavelength regions of the e-m spectrum. However, the fact that they are a sampling tool means that the resulting information has its limitations. A much more detailed analysis of the radiation from stars comes from their spectra.

Sorting Out Stellar Spectra

By the beginning of the 20th century it was realized that the spectra of many stars showed absorption lines due to hydrogen, the simplest of all chemical elements. So the first attempted system for classifying stellar spectra was based on the prominence of these hydrogen lines. The idea was to designate stars with the darkest hydrogen lines as being of spectral class "A," those with the next strongest lines of class "B," and so on. However, developments in physics, particularly quantum mechanics, together with the increasing recognition that the chemical composition of many stars was broadly similar (i.e., a lot of hydrogen, a fair amount of helium, and essentially traces of many other elements such as calcium, sodium, and iron) made it evidently clear that what principally determined the appearance of a star's spectrum was the temperature of its surface layers.

In addition the idea behind classifying the spectra of stars was to have some kind of systematic sequence, rather than simply a "pigeon holing" scheme, so what started off as a hydrogen line sequence became a temperature sequence, running smoothly from the hottest stars through to the coolest ones. This is why the spectral classification of stars is often referred to as the *spectral sequence.*

Much of this work was done at Harvard College Observatory, particularly by Annie Jump Cannon. The slightly unfortunate thing was that the initial orderly sequence of letters "A," "B," "C," etc., used to denote hydrogen line strengths, got a bit messed up when the system converted to a temperature sequence. Not only did the order of the letters change, but some letters were actually dropped. Spectral class "A" stars have the strongest hydrogen lines, but they now fall in after classes "O" and "B" because they are cooler.

Anyway, all astronomers both amateur and professional soon get to learn the letters of the spectral sequence, "O B A F G K M," with the help of the famous mnemonic "Oh Be A Fine Girl Kiss Me." Stars become cooler then as we move through the spectral sequence from class "O" to class "M," and their visual spectra become more complex with the appearance of increasing numbers of absorption lines as the temperature falls. Indeed, the sheer complexity of stellar spectra necessitated subdividing the letter categories using numbers. So for example the spectral class of the Rigel is B8, Arcturus is K2, Vega is A0, and so on, and incidentally, the very hottest stars come in at class O5, not O0. Just on a matter of terminology, stars are very often referred to in the literature as being "early" or "late" – this has nothing whatsoever to do with a star's "age," but whether the star's spectral class puts it earlier or later within the spectral sequence, so that classes O, B, and A would be referred to as "early," whereas classes K and M would be referred to as "late." Indeed, these terms are often applied even within a single spectral class, so that a G2 star would be "earlier" than a G8 star, and so on.

So like the B – V color index, the spectral sequence is a temperature scale for stars, which means that the hotter early-type stars should be the most luminous. Luminosity is another of those very important numbers for stars, and clearly any relation between luminosity and temperature is bound to be an astronomer's "crowd pleaser." Once the spectral sequence had been established the next step was to investigate the relationship between a star's spectral class (i.e., its temperature) and its luminosity. Enter Messrs. Hertzsprung and Russell.

The Hertzsprung–Russell Diagram

By the early years of the 20th century a reasonable sample of stars had had their distances and hence their absolute magnitudes (on the traditional visual magnitude scale) determined. This enabled the Danish astronomer Ejnar Hertzsprung and the American astronomer Henry Norris Russell to independently (though around the same time) investigate the relationship between luminosity and spectral class. The resulting Hertzsprung–Russell (or HR) diagram, well known as the most important diagram in all of stellar astronomy, is thus effectively a plot of the luminosity of stars against their temperature. By convention luminosity, which is represented by absolute magnitude, runs up the vertical axis, so that while the most luminous stars appear at the top of the diagram, because the magnitude scale runs backward, the numbers on the vertical axis run from top to bottom. In addition, spectral class (in effect, temperature), which is plotted along the horizontal axis, also runs backward, so that the hottest stars are on the left of the diagram (all perhaps a little perplexing for non-astronomers and probably quite bewildering for astronomers, if it were done any other way).

Since Hertzsprung's and Russell's time more data has become available, enabling more points to be added to the HR diagram, shown in its most basic form in Fig. 5. Its most striking feature is that the positions of stars are not distributed randomly over the diagram, but instead fall within very distinct regions, with some regions being more heavily populated than others. Most stars fall within a region that runs from the top left corner, i.e., the hot luminous region of the diagram to the lower right, or cool, dim region. This group of stars, which includes our own Sun, forms the *main sequence*, and it is in every sense a visible illustration of the spectral sequence itself; furthermore, it clearly shows that as (approximate) blackbodies, the hotter stars give off more energy and are thus more luminous. Have a look back at Figs. 2a and 3. It is no coincidence that a plot of the log of the temperature, which is directly related to the emitted flux against B – V (and thus a measure of a star's color), resembles the main sequence on the HR diagram.

The B – V color index for a star comes from making two relatively simple observations of the star, and the B – V scale forms a continuous running series of numbers. By contrast, positioning a star in its correct place on the HR diagram is a much more difficult process, which involves two separate quantities – the luminosity and the spectral type (astronomers call this a *two-dimensional classification scheme* as opposed to the B – V scale, which is a one-dimensional scheme).

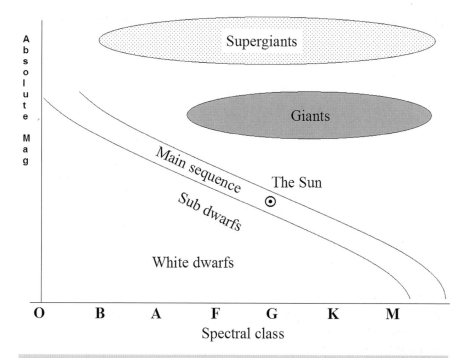

Figure 5. The most basic form of the Hertzsprung–Russell diagram showing the main population zones for stars.

Fixing the spectral type involves the careful identification and measuring of the relative intensity of perhaps a large number of absorption lines in the spectrum. The star then has to be placed in what amounts to the appropriate predetermined "slot" in the spectral sequence. The luminosity is represented perhaps not surprisingly by the star's absolute magnitude, but unless there is some way to either determine or at least make a reasonable estimate of the star's distance, or alternatively some other method of estimating its luminosity, then "that" as they say "is that." This can be particularly difficult for the most luminous stars, which are for the most part very distant and in fact only visible simply because they are very luminous. Even so, once sufficient data have been acquired the resulting HR diagram can be used as a kind of template against which other stars can be compared and hopefully, as a result, reveal important information about themselves.

In addition to the main sequence, the next most striking feature of the diagram is that at the high-temperature end, there is a fairly narrow range of absolute magnitudes (except for the white dwarf stars, which are very much a "special case") and thus a corresponding low range of luminosities.

By contrast, the coolest stars appear to belong to either the low-luminosity end of the main sequence or form a group of highly luminous stars at the top right of the diagram. As approximate blackbodies, cool red stars will emit less radiation per square meter of their surface layers than hotter stars. So for a cool red star to be luminous, it must have a large surface area or, in other words, it must be a giant star – a red giant. So just as the HR diagram classifies stars according to their spectra, it will inevitably also categorize stars according to their luminosity, and as we move toward the cool end of the diagram, luminosity shows itself ever increasingly in terms of a star's physical size. This means that there are stars that can be called "giants" and those that are called "dwarfs."

Suppose you are studying the spectra of two red stars, and you decide that both their spectra are of, say, class M3. You don't have any information on their absolute magnitudes, so is there any other way to tell if either star is a red dwarf or a red giant? The answer is "yes." Although stars can vary greatly in both luminosity and size, the masses of stars do not cover such a large range. What's more, it is known that giant stars have most of their mass concentrated at the center, so this means that the outer layers of giant stars are of very low density (a red giant star has often been referred to as a "red hot vacuum").

When absorption lines are produced in a gas that is relatively (but not too) dense, such as is the case for a red dwarf star, there is a spectroscopic effect called *pressure broadening*, which literally causes absorption lines to be broader than they would be if produced in a less dense gas, such as the outer layers of a red giant. So a telltale sign of a red giant star is that the absorption lines in its spectrum are narrow. The low-density regime of a red giant also favors the appearance of more actual absorption lines from the heavier elements which have multi-electron atoms. These spectroscopic differences even cause the B – V color index to be slightly larger for a red giant than that of a red dwarf of identical spectral class. This is another way of saying that red giants are redder than red dwarfs.

So the HR diagram, which came out of the spectral sequence, itself gave rise to a luminosity classification scheme for stars. The scheme, which was developed by Morgan, Keenan, and Kellman at the Yerkes Observatory, is often referred to as the MKK system, and its significance makes more sense when the luminosity classes of stars (see below) are effectively plotted in the form of an HR diagram, as shown in Fig. 6. The scheme divides the HR diagram into a series of zones that clearly become wider with lower stellar temperatures. Unfortunately, there doesn't seem to be a catchy mnemonic for remembering luminosity classes, and they are designated by ancient Roman numerals, which doesn't help. These then are the official stellar luminosity classes:

Luminosity classes

Ia	Most luminous supergiants
Ib	Less luminous supergiants
II	Bright giants
III	Normal giants
IV	Subgiants
V	Main sequence (dwarfs)
VI	Subdwarfs
VII	White dwarfs

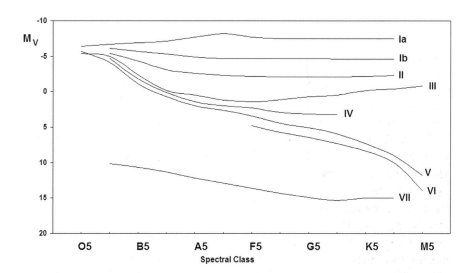

Figure 6. The MKK stellar luminosity classes, plotted according to absolute magnitude and spectral class.

So now an "ID card" for a star would include the luminosity class as well as the spectral class; for example, the Sun is classified as G2 V – a main sequence star of spectral type G2; Betelgeuse, on the other hand, comes in at M2 Ia – a mighty big but relatively cool star.

A final word here might be to note that stars on the main sequence, which of course includes our own Sun, are both classed and referred to as "dwarf stars." When thinking of dwarf stars, we tend to think of maybe white dwarfs or perhaps red dwarfs, but here the term "dwarf" is simply used to distinguish main sequence stars from the giants. Any star whose spectrum and luminosity class place it in one unequivocal position on the HR diagram can be regarded as "normal," and many stars (usually the most interesting ones) by this benchmark are not normal.

Using the HR Diagram "Template"

The spectrum of a star together with more subtle features that distinguish the star as either a dwarf or a giant fixes its position on the HR diagram. The luminosity that is derived for the star can, in theory, produce two very important numbers, or "parameters," for the star. The spectrum itself is a measure of the star's effective temperature, just like the B − V color index, and, as we have seen, by approximating the star to a perfect blackbody, the total flux emitted by each square meter of the star's surface layers is given by the Stefan–Boltzmann law. The star's luminosity is just equal to this value multiplied by the star's surface area, which itself is simply equal to $4\pi R^2$, R being the star's radius. So the radius of the star can be determined.

Secondly, the luminosity – i.e., the total flux emitted by the star – can be used to determine how much of this flux would be received at a distance of 10 pc, giving us in turn the star's absolute magnitude. We can easily measure the star's apparent magnitude, and a simple application of the distance modulus formula then gives us the distance to the star. A distance measured in this way is called a spectroscopic parallax (there is no "parallax" as such involved, but the name comes out of the trigonometric parallax method for measuring star distances). Stellar parameters derived using the HR diagram can be subject to possibly large errors; nonetheless, this diagram is clearly a very useful tool.

A Different Kind of HR Diagram

It's clear now that both the spectral type and the B – V color index are measures of a star's temperature, and, as we've seen, classifying a star according to its spectral type is a harder job than determining its B – V color index. So, instead of plotting absolute magnitude against spectral type, it's much easier to plot absolute magnitude against B – V, and what we have then is usually referred to as an *observational HR diagram*. We still have the same kind of distribution of stars on the diagram – the main sequence, the giants, and the dwarfs; but we now have stars classified according to their luminosity as well as simply by their B – V values. When astronomers use "normal stars" to establish a relationship between B – V and effective temperature, they tend to use main sequence stars.

Conclusion

Our original aim in this chapter was to "take the temperature" of a star. At first this seemed to be very straightforward, mainly because the equations that we used to do the calculations are very simple. We see, however, that there's a bit more to it than just plugging numbers into a formula, but at least on the way we've hopefully learned a lot more about the stellar magnitude system and why those special photometric filters are so useful.

Key Points

- The continuum part of a star's spectrum resembles a blackbody spectrum.

- The temperature of a star, determined using the Stefan–Boltzmann equation and which assumes that the star radiates as a perfect blackbody, is called the effective temperature, and this is what is almost always meant by "the temperature of a star."

- The use of photography meant introducing the photographic magnitude m_{pg} in addition to the original visual magnitude for every star.

- The use of special photographic emulsions and a special filter resulted in the photo-visual magnitude m_{pv}, which closely approximated the visual magnitude itself.

- The formula $m_{pg} - m_{pv}$ defines the international color index for a star, and it is a measure of a star's temperature.

- The development of photoelectric photometry resulted in the introduction of many so-called photometric systems, of which the most well known is the UBV system of Johnson and Morgan.

- The UBV system results in a series of color indices, which are determined by taking the difference between star magnitudes as measured using special filters, each of which covers a specific range of wavelengths.

- The star Vega is defined to have all color indices equal to 0.0 and as such defines the zero point for all magnitude scales in the Johnson and Morgan system.

- A great deal of work has enabled astronomers to establish a relationship between the $B - V$ color indices for "normal stars" and their effective temperatures.

- A star's magnitude that results from determining the total flux received from it across the entire e-m spectrum is called the apparent bolometric magnitude.

- A plot of luminosity or absolute magnitude against stellar spectral class results in the famous Hertzsprung–Russell diagram, which itself showed that stars also need to be classified according to their luminosity.

The Photons Must Get Through – Radiative Transfer

If starlight shone clear and uninterrupted across light years of space and shone just as clear down through Earth's atmosphere into our telescopes, life for both professional and amateur astronomers would be much simpler, though possibly nowhere near as interesting. However, a great deal can and does happen to our river of starlight before it reaches our telescopes and detectors. In fact, even before that starlight can leave the photosphere it has to get from its source – the nuclear reactor core at the heart of the star – to the star's surface layers. This is a relatively short journey in terms of distance but by far the longest in terms of time, and much depends on the very structure and composition of the star itself.

Explaining and describing in detail what goes on between star and telescope forms a big part of both observational and theoretical stellar astronomy, and it can involve some pretty grim mathematics, which is why, perhaps, professional astronomers get paid those big bucks. However, as amateurs we can still gain a truly fascinating insight into what is known as radiative transfer, or sometimes radiative transport, with the ever important bit of basic physics and maybe just the odd equation, which as before will enable us to easily calculate some important numbers. These numbers will in turn give us a better feel for what's going on. Oh yes! And this is where we'll need that number "e," or 2.718.

K. Robinson, *Starlight*, Patrick Moore's Practical Astronomy Series,
DOI 10.1007/978-1-4419-0708-0_6, © Springer Science+Business Media, LLC 2009

Absorption – The Photons Get Taken Out

To the astronomer, the most (all too) familiar thing that can happen to starlight is that it can be obscured by clouds. Clearly the tiny water droplets that make up a bank of cloud form a medium that is pretty opaque to visible wavelength photons. However, to some wavelengths, for example, in the radio region of the e-m spectrum, this bank of cloud presents no problem, and the radio photons pass through what to them is a transparent medium. So, the first thing to be very clear about here is that when starlight passes through some intervening medium such as interstellar gas or dust, the effect of this medium will be different and sometimes very different for the different wavelengths that make up the starlight. Thus, from now on in this chapter, when we use the terms "light" or "starlight" we mean light that consists of only a very narrow range of wavelengths (remember, we said previously that light of a single wavelength is in practice not feasible), though we are free, of course, to change from one narrow wavelength range to a different one in another part of the spectrum.

Let's start then with a situation where we have starlight shining through a region of space that is occupied by some absorbing medium such as a nebula made of gas and/or dust. We have to be very specific here about the meaning of the word "absorption"; absorption means that a photon is completely lost to the beam of starlight. There are various ways in which this can happen. For example, the photon may *ionize* a gas atom by removing one of its outer electrons, or the photon's energy may serve to warm up ever so slightly a grain of interstellar dust. For whatever reason, the photon is lost, and because a vast number of such events take place within such an absorbing medium, a great many photons get taken out, and so the intensity of the light from the star is diminished. By how much is it diminished, though?

To keep things simple we need to "pin down" this cloud of material by first of all giving it a finite thickness or depth of say "*s*" ("*s*" could be in light years or kilometers or whatever). The next thing is to insist that the density of the material is the same everywhere; in other words, wherever we are in the cloud, there are the same number of kilograms of material per cubic meter (in a real nebula there would in fact only be a very tiny fraction of 1 kg/m^3). It also means, for example, that the cloud or nebula has no fuzzy edges but a sharply defined boundary at both the entrance and the exit faces for the beam of starlight.

Finally, the composition of the nebular material must be the same everywhere, so that if the nebula is a mixture of gas and dust, it must be the same uniform mixture of the same kind of gas and the same kind of dust. Clearly a nebula in which the density and/or composition vary from place to place is capable of affecting an incoming beam of starlight in an endless number of ways, so it really is vital here to keep things simple so that we can more easily understand what happens to the photons. Thus, we must have a completely uniform slab-shaped nebula of a certain thickness or depth through which a beam of starlight passes.

The first thing that should be fairly obvious about our "standardized" nebula is that the greater the value of "s" the more the starlight will be diminished, simply because the greater the distance the light has to travel through the nebula, the more atoms/dust grains there are to intercept the photons. This situation also works the other way around, in that a more intense beam of starlight means that there are more photons in the beam, which means that to an individual atom/grain there is a greater chance of encountering one or more photons; again, more photons get taken out. So the greater the distance traveled through the nebula and also the more intense the incident beam of starlight, the greater the actual drop in the intensity of the beam of starlight that emerges from the nebula. Because we have "standardized" our nebula, its chemical/physical composition can be ignored here, but in a real situation this can become the dominant factor.

The final thing to say here is that even within our simple nebula, there's a major element of chance about things. It would be difficult enough to follow the fortunes of one individual photon as it enters the nebula; it may get absorbed or it may survive to emerge from the nebula's far side. But to follow the individual fates of a vast number of photons is clearly impossible. We are, in fact, at the mercy of those wretched things called statistics. In crude terms this means that if "x" photons enter the nebula today and "y" photons emerge, then tomorrow the number of photons emerging from an initial population of "x" photons will probably be very close to "y" but not exactly the same as "y." This is actually not something that we really need to worry about too much; we simply need to be mindful of the fact that when we talk about the numbers of photons that get absorbed or which survive, we are dealing with vast numbers of photons anyway, which may vary ever so slightly from theoretically calculated values.

A population of starlight photons enters the nebula, and let's think about what's happening as they progress through it. If we know the initial number of photons (in principle this can actually be calculated by simply dividing the total energy of the starlight, i.e., its intensity, by the energy per photon) and if the same number of photons were absorbed in each succeeding meter of nebula, then knowing the total depth of the nebula ("*s*" in meters), it should be very simple to work out the number of photons that survive to emerge from the nebula.

However, things don't work this way, because right from the start, photons are absorbed, and so the number remaining also starts to drop immediately, which means that further into the nebula there are progressively less photons to be "grabbed" by a gas atom or a dust grain, and so the number of absorptions also progressively drops. So rather than thinking about the actual number of photons that are absorbed in, say, the first, second, third, etc., meter of the nebula, think instead about the *fraction* of the photons absorbed in each successive meter. For an initially large incident number of photons, "some given fraction" will mean a relatively large number of actual photons are absorbed, whereas a progressively decreasing population would result in a smaller number of photons being absorbed as the light progresses through the nebula. This is exactly the kind of behavior we're looking for; it means that for our standardized nebula the *fraction*, rather than the actual number of photons, absorbed *stays the same for each succeeding meter*, and because the population of photons falls off as we move through the nebula, the actual number of absorptions also falls off. The key then is the fractional drop in the number of photons per meter (or it could in fact be any small unit distance) of nebula.

This fractional drop in the number of photons for a small unit or fixed distance in an absorbing medium is called perhaps rather loosely the *absorption coefficient* of the medium. More rigorously it is called the *linear absorption coefficient*, and it is almost always represented in the literature by the lower case Greek letter kappa or "κ." The actual value of κ depends on the density and composition of the absorbing material, and clearly a higher value for κ means more photons are absorbed and the medium is relatively opaque, whereas a lower value means that the medium is more transparent. Finally, just as a reminder, you'll often see the absorption coefficient written as κ_λ in order to emphasize the fact that its value is more often than not dependant on the wavelength of the light being absorbed.

Here's a very simple (though very unrealistic) example to show how the linear absorption coefficient works. Let's say we start with a population of 100 photons and a nebula 20 m in depth. Now suppose that in the first meter of the nebula, 20 photons are absorbed. If 20 photons were absorbed in each succeeding meter, then clearly the supply of photons would be exhausted after 5 m ($5 \times 20 = 100$), leaving no surviving photons to emerge from the nebula. However the fractional drop in the number of photons, i.e., the linear absorption coefficient in this case is 20/100, which equals 1/5 or 0.2, and we make the bold assumption that for our standardized nebula this fraction (the value of κ) remains the same, or constant for each of the 20 m through the depth of the nebula.

So, after the first meter 0.2×100, 20 photons are absorbed, which means that 80 remain. In the second meter 0.2×80, 16 photons are absorbed, leaving 64 to enter the third meter and so on. See if you can work out how many photons are absorbed in each of the succeeding meters of the nebula, until you've worked out how many emerge from the nebula's far side. You'll see that the numbers of photons entering and emerging from successive meters are now no longer whole numbers. Don't worry about this; remember this is an unrealistic example. In reality, there would be a vast population of photons, and the chance element mentioned earlier means that while the numbers of photons involved might not agree precisely with the calculated values, they would in fact be very close, and of course they would be whole numbers.

The numbers of photons that survive after each successive meter of our simple nebula are listed in Table 1.

Table 1. One hundred photons enter a nebula that has a linear absorption coefficient "κ" equal to 0.2. As we can see, just over one photon makes it to the other side. In reality, the nebula would be vastly bigger, but the number of photons would also be vastly greater.

No. of meters into nebula	No. of remaining photons
0	100
1	80
2	64
3	51.2
4	40.96
5	32.77
6	26.22
7	21
8	16.8
9	13.44
10	10.75
11	8.6
12	6.88

Table 1. (Continued)

No. of meters into nebula	No. of remaining photons
13	5.5
14	4.4
15	3.52
16	2.82
17	2.26
18	1.81
19	1.45
20	1.16

On the face of it then, determining how the intensity of the starlight diminishes as it progresses through our standardized nebula doesn't seem too difficult using the method just described. It gets more difficult when we try to calculate the number of photons that remain after, say, 0.5 m, or 5.83 m, or indeed any arbitrary distance into the nebula rather than a whole number of meters. It is possible with the aid of a calculator and if your math is up to it you should have no difficulty in working out how it's done. However, we won't go into the details, because the professionals don't do it this way, and one of the main reasons for this is that you really do run into trouble if, as is often the case, the value of κ varies as we move through the nebula. This results from changing conditions within the nebular material, such as changes in the density.

So here's how the professionals do it. For the moment we'll stick with our simple standardized nebula example with an incident beam of 100 photons and a constant linear absorption coefficient of $\kappa = 0.2$ m^{-1}, as before. Again, as before, we'll consider the situation after 1, 2, 3, etc., meters into the nebula until we reach the far side. This time we take the number of meters into the nebula and multiply this distance by the value of κ to give us 0.2, 0.4, 0.6, etc., and then take the negative of each of these numbers to get –0.2, –0.4, –0.6, and so on.

Now use your pocket calculator and the "x^y" key to raise that number "e" or 2.718 to the power of each of these negative numbers. In other words, calculate $e^{-0.2}$, $e^{-0.4}$, $e^{-0.6}$, and so on. Many of you will know that "e" is one of those recurring decimals (just like "π"). If you prefer to use a more accurate value then enter the number "1" on your calculator; now press the button marked "INV" for "invert," followed by the button marked "Ln" (this stands for natural or sometimes Naperian logarithm –

not to be confused with common logarithms that employ the use of the button marked "LOG"). This will give you a more precise value, such as 2.718281828, but for now just 2.718 will do. Finally, multiply each of the resulting 20 numbers by 100 – the number of photons that initially entered the nebula. The results are listed in Table 2.

Table 2. Here we have calculated the number of remaining photons, using the number "e" or 2.718. This is how the professionals do this kind of thing, and while the results are not quite the same as before, they are similar and in fact in a real situation involving a very large nebula and vast numbers of photons, the results would be virtually identical.

No. of meters into nebula	No. of remaining photons
0	100
1	81.87
2	67.03
3	54.88
4	44.94
5	36.79
6	30.12
7	24.66
8	20.19
9	16.53
10	13.54
11	11.08
12	9.07
13	7.43
14	6.08
15	4.98
16	4.08
17	3.34
18	2.73
19	2.24
20	1.83

These 20 numbers represent the number of photons that remain after the first, second, third, etc., meter into our nebula. When compared to the corresponding values we calculated by the previous method, we can see that while they are roughly in the same ball park, they are not exactly the same. However, as we admitted at the start, our example here is very unrealistic; a more realistic situation would involve vast numbers of photons and a nebula that was many, many millions of meters in depth. In this case there would be essentially no difference between the two sets of corresponding numbers.

Just as a final illustration, we've plotted the number of surviving photons as we progress meter by meter through our nebula, using both methods as shown in Fig. 1.

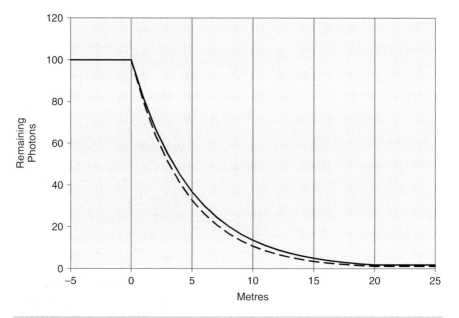

Figure 1. This shows how the number of photons decreases as they pass through a "nebula." The *dashed plot* uses the results of our first method for calculating the number of remaining photons. The *solid plot* employs the "professional method," using the number "e." In a realistic situation, these two plots would be indistinguishable, but the first method of calculation would be much more laborious.

The two plots have almost the same shape, and indeed with more real-istic parameters, they would be virtually indistinguishable. The really important point here is that the plot shows just how the intensity of starlight falls off as it passes through an absorbing medium.

What probably is not so convincing at this stage is why use the num-ber "e" raised to – of all things – a power that is a negative number, to calculate the remaining number of photons, rather than the first method, which perhaps seems more intuitively logical. The second method is actu-ally entirely equivalent to and is in fact more mathematically rigorous than the first method, but is arrived at by using calculus, which puts it beyond the scope of this book.

Suffice it to say that for a professional astronomer with sufficient mathematical skills, perhaps the aid of a powerful computer, and if it's a good day and the value of κ varies in a nice, regular way, then the second method using "e" delivers the goods. Most importantly for us, though, this second method for calculating the intensity of starlight as it passes through an absorbing medium introduces us to something very important in radiative transfer. In order to calculate the intensity, we first had to multiply the absorption coefficient κ by the distance or depth into the nebula. If κ is large, then we don't have to go very far into the nebula before the absorption becomes significant.

Conversely, for small values of κ, we have to go to greater depths for there to be significant absorption, so clearly what ultimately determines how much the intensity of starlight drops is the product of the absorp-tion coefficient and the distance into the nebula, i.e., $\kappa \times s$. (The "s" here represents distance into the nebula rather than the entire depth of the nebula.) This all-important product is called the *optical depth* and is invariably represented in the literature by yet another Greek letter, this time the letter "tau" or τ; note once more that you'll often see it writ-ten as τ_λ to emphasize its wavelength dependence. Figure 2 illustrates the basic idea behind optical depth. Put simply, a large optical depth means that much of the incident light is absorbed, whereas a small optical depth results in most of the light making it through to the other side.

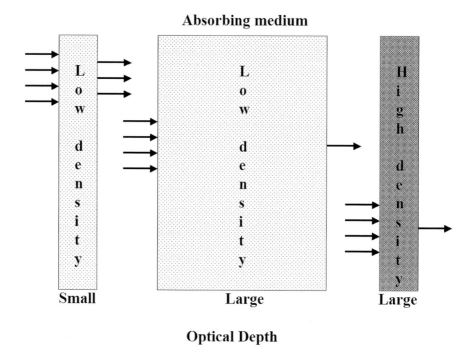

Figure 2. The optical depth in an absorbing medium depends on the linear absorption coefficient κ and the distance traveled through the medium.

Now let's catch our breath by being a bit more specific about something that, until now, we've been rather casual. When talking about starlight entering a nebula we've talked about numbers of photons, though sometimes we've used the term intensity, and in the interests of keeping things simple we've tacitly assumed that starlight both enters and emerges from the nebula, traveling in a direction that is at right angles or normal to the face of the nebula.

In chapter *Space – The Great Radiation Field* we saw that the intensity measures the rate of flow of radiation, which is confined to what amounts to a tiny range of directions, and we made the point that it does not change with distance. Thus, when talking about the way in which starlight is affected as it passes through what may be a very large nebula we should indeed talk in terms of its intensity; using the flux that we know does change with distance would add an extra complication. However, there is an important situation involving the absorption of starlight over such a short distance. This is the passage of starlight through Earth's atmosphere, which we shall look at in chapter *In the Space Between Stars*.

So if we now call the intensity of starlight that enters the nebula "I_{in}" and that which emerges "I_{out}," we're talking about light that is traveling in one specific direction. A very important consequence of this is that if our beam of starlight enters the slab-shaped nebula at an angle other than 90° to the nebula's face. Then, by the time it emerges from the far side, it will have traveled a greater distance. So the optical depth and hence the drop in intensity through our nebula is greater along directions that are not "normal" to the nebula's face, as shown in Fig. 3; we'll see the importance of this when we look at the way in which light travels through the outer layers of a star and then through Earth's atmosphere.

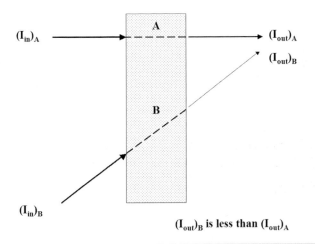

$(I_{out})_B$ is less than $(I_{out})_A$

Figure 3. In an absorbing medium (here, for example, we could have the photosphere of the Sun) with a fixed linear absorption coefficient, the optical depth depends on the actual distance traveled through the medium. So the optical depth along path "B" is greater than that along path "A," which means that for the same incident intensity, the emergent intensity along path B is less than that along path A.

Finally we can now be a bit more formal about the "professional method" for calculating the behavior of starlight as it passes through an absorbing medium. Calling the initial or incident intensity "I_{in}," the linear absorption coefficient "κ" multiplied by the depth "s" into the medium gives us the optical depth "τ," and the emergent intensity "I_{out}" is given by

$$I_{out} = I_{in} \times e^{-\tau} \tag{1}$$

There's a general convention or, if you wish, a kind of demarcation boundary regarding the value of the optical depth "τ"; if τ is less than 1, then the medium is said to be *optically thin*, whereas for optical depths of greater than 1 the medium is *optically thick*. For a value of $\tau = 1$, $e^{-\tau}$ is equal to 0.368 to three places of decimals, and so for an absorbing medium whose optical depth puts it on the borderline between being optically thick and optically thin, the intensity of the emergent starlight is reduced to about one third of that which enters the nebula. Having said all this, there is no law saying that starlight, which enters a nebula traveling in one specific direction, must emerge from the nebula traveling in the same direction. This has very interesting consequences, as we'll now see.

The Photon Gets a Second Chance (and Maybe More) – Scattering

As emphasized above, absorption involves the effective destruction of photons in our beam of starlight. The photons are lost because their energy has been used to do something else, such as warming up dust grains. However, it is possible for some photons to be removed from the beam of starlight but yet survive. The photons are effectively removed because they are simply no longer traveling in their original direction, and this happens because the photons have been *scattered* – sent traveling in other directions by the gas atoms or dust grains rather than absorbed.

It isn't necessary here to go into the actual mechanisms that cause scattering; what is important is that the process of scattering is conspiring with absorption to diminish the intensity of the starlight that we observe; indeed, just as we introduced the idea of the linear absorption coefficient as a measure of the relative drop in the intensity for each small unit distance through the absorbing medium, we can also define the *scattering coefficient* (often denoted by the Greek letter sigma or "σ") in exactly the same way, except that this time the relative drop in intensity is due to scattering processes. The linear absorption coefficient and the scattering coefficient can simply be added together to give what is called the *extinction coefficient* (this time denoted by the Greek letter chi, or "χ"), so that

$$\chi = \kappa + \sigma \tag{2}$$

The extinction coefficient is sometimes loosely referred to as the *opacity* of the medium. The optical depth for a distance "s" into the nebula is now given by $\tau = \chi \times s$, and the formula for the emergent intensity is the same as Equation (<eref 1), except that the optical depth now results from both absorption *and* scattering.

As with the absorption coefficient, the value of the scattering coefficient depends on the wavelength of the light, but for gas atoms the degree of scattering is proportional to the reciprocal of the fourth power of the wavelength of the light, i.e., proportional to $1/\lambda^4$. In other words, if you halve the wavelength of light the amount of scattering increases by 16 *times*. This results in blue light being very much more susceptible to scattering than red light, which in turn renders our daytime sky blue, and as we shall see does interesting things to starlight.

By contrast, infrared photons suffer relatively little scattering, which results in the infrared astronomer being able to "see" much greater distances and through clouds of interstellar dust.

Scattering also plays a "double agent's game" in radiative transfer; just as photons can be removed from our beam of starlight and scattered into other directions, clearly photons, which initially enter the absorbing medium traveling in some arbitrary direction, can find themselves being scattered in our direction, and so our beam of starlight gets reinforced by these imported photons.

It may be that in some situations the additional drop in intensity caused by scattering is more or less matched by scattering into the beam from other directions, and an obvious example where scattering comes into its own is in so-called reflection nebulae (a more appropriate name for them might indeed be "scattering nebulae") such as NGC 1977 in Orion. We observe this nebula because of light from neighboring stars, initially traveling in some other direction, which has been scattered and combined into a beam that travels toward us. One thing to note here is that some of these scattered photons may themselves suffer further scattering, or even absorption, before they reach the side of the nebula nearest to Earth.

New Photons for Old – Emission

A good question to ask at this point is when a photon is destroyed, where does its energy go? We've already mentioned that a visible light photon can help to warm up a dust grain, for example; the consequence of this is that the dust grain itself begins to emit thermal radiation – photons. Because of the dust grain's relatively low temperature, these emitted photons will be low energy ones in the infrared part of the spectrum.

What has happened here is that a visible light photon of relatively high energy has been converted or degraded into several lower energy photons. Now, if you are an optical astronomer, you would say that the light from a star that sits behind this cloud of obscuring material has been diminished, and you might go on to try and work out the optical depth of the material at visible wavelengths. On the other hand if you happen to be an infrared astronomer (some amateur variable star astronomers now do infrared photometry), your knowledge of the star's other properties tells you that given its temperature it should emit a certain level of radiation at infrared wavelengths. What you observe, though, is extra infrared radiation, which is due to emission from the surrounding dust. This, as previously mentioned in relation to Vega, is called an *infrared excess*, and it means that the optical astronomer's loss is the infrared astronomer's gain; it is also, of course, the telltale signature that the star you are observing may be surrounded by dusty material.

All is not lost for the optical astronomer, though, because this process of photon degrading and recycling goes on right across the electromagnetic spectrum, with results that are often spectacular and much loved by observers of deep sky objects. Just as dust grains can convert optical photons into infrared photons (i.e., extra infrared emission), high-energy ultraviolet photons can be converted into visible light photons as a result of being absorbed by gas atoms. The energy of an ultraviolet photon is used to ionize, or remove an electron from, a gas atom.

This is, in every sense of the word, a version of the photoelectric effect, which we encountered in chapter *From Light to Starlight*. In this case there is, of course, no "wire" to lead the electrons away in the form of an electric current. Instead the electrons are free to move around until, as inevitably happens, they encounter another gas atom, which has itself lost one or more electrons by this process of *photoionization*. The positively charged atom captures the negatively charged electron, which as a result loses energy. However this energy, which is equivalent to that of the original ultraviolet photon, which removed the electron from its original parent atom, is lost in stages as the electron drops down through its new atom's energy levels. This *cascade process*, as it is often called, causes the atom to emit a series of lower energy, i.e., visible light photons, which because of the well-defined energy level structure of atoms results in a series of emission lines with well-defined wavelengths.

If the gas is some distance away from the source of the ultraviolet photons, then we "see" a more or less pure emission spectrum, as is the case, for example, with planetary nebulae. If, on the other hand, the absorbing gas forms a close proximity shell or layer around the star, then we may see bright emission lines superimposed on the star's otherwise relatively normal spectrum. Because these lines result essentially from the recombination of electrons with atoms, they are called *recombination lines*. There are in addition other processes going on in nebulae that also result in emission lines. For more details on this you might want to have a look at this author's *Spectroscopy – The Key to the Stars* (published by Springer).

So, we see now, that we can have a situation where extra emission comes from the absorbing medium itself; this medium may be in the form of a nebula (i.e., low-density gas), but it could also be a layer within a star, which absorbs high-energy photons from the deeper and hotter layers and recycles them in the form of lower energy photons, which then head toward the star's photosphere.

The difference here is that the stellar material forms a relatively dense medium, which produces a continuous spectrum rather than an emission-line spectrum, which we observe in the case of a nebula. What's more, just as starlight entering some medium suffers absorption, so, too, will light emitted by the medium itself. The medium, in fact, acts as an absorbing medium for its own emission.

Clearly, emission from the far side of this material will suffer more absorption as it heads in our direction than that which is produced closer to our side. If we focus on one small region of our absorbing medium, say, a one-meter cube, then the quantity of radiation being emitted by this small volume element will first of all depend on what the gas is made of. Atoms that are more likely to be photoionized will result in more emission; also, the denser the gas, the more atoms there are, and again the more emission there will be.

We can also assume that the radiation emitted from within our one-meter cube spreads outward from the cube equally in all directions, i.e., it is isotropic, so that the same quantity of energy, or the *intensity* in every direction, is the same. This quantity of energy is called the *emission coefficient* or sometimes the *emissivity*; it's usually represented by the Greek letter epsilon, or "ε," and as you'd expect it is wavelength dependent and so is often written ε_λ."

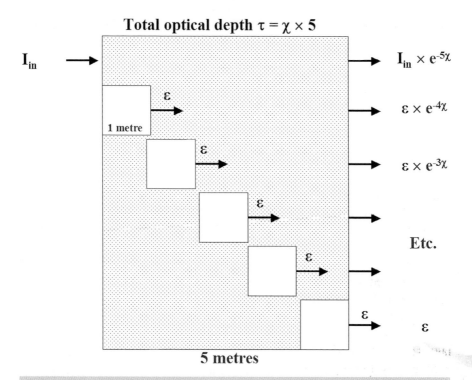

Figure 4. Here we have a simple slab-shaped nebula of constant extinction coefficient and emission coefficient, divided into one-meter cubes. Emission from each cube will suffer absorption within the nebula, depending on how much material lies between it and the nebula's near side. The resulting emission seen coming from the nebula is the sum of the contributions from all the cubes, plus any unabsorbed or scattered emission from beyond the nebula itself.

Think once again about a simple slab-shaped nebula, which has a depth of 5 m and has a constant extinction coefficient "χ" (combining both true absorption and scattering) of 0.2 m^{-1}. We can now imagine 5 one-meter cubes lined up and running from the far side of the nebula to the side nearest to us, as shown in Fig. 4. Each of these one-meter cubes emits radiation of an initial intensity ε that travels toward us. The light that travels from the one-meter cube on the far side has to pass through the intervening 4 m of nebula and so, adapting Equation (1) the intensity of radiation from this cube that emerges from the near side of the nebula will be given by

$$\varepsilon_{out} = \varepsilon \times e^{-0.2 \times 4} \tag{3}$$

In this case 0.2×4 is the optical depth through the 4 m of nebula. In turn, the emergent intensity from the next cube in will be equal to

$$\varepsilon_{out} = \varepsilon \times e^{-0.2 \times 3} \tag{4}$$

and so on. Note that ε_{out} here is different in each case because the optical depth is dropping as we progress from the far side to the near side of the nebula.

Finally, emission from the cube at the nearby face emerges more or less unhindered. To calculate the total emission from our nebula we simply add together all the separate values of ε_{out}. However, once again, in order to do this calculation rigorously with a realistic nebula, we would have to use calculus. The resulting equation for the intensity of emission emerging from the nebula is at least fairly simple (it is included here just to show what the actual result is), i.e.,

$$\varepsilon_{out} = \frac{\varepsilon}{\chi} \times (1 - e^{-\tau}) \tag{5}$$

and "τ" here is just the optical depth through the whole nebula. So this is what we "see" of starlight that has been recycled within the nebula into the wavelength we are observing and which emerges from the nebula traveling in our direction. Notice that this equation involves the quantity "ε/χ," i.e., the ratio of the emission coefficient to the extinction coefficient or, in other words, the ratio of the emission to absorption within each small unit volume element in the nebula. This is an important quantity in the theory of radiative transfer and is given a special name – the *source function* or "S." A relatively large value for S simply tells us that at a given wavelength each small volume element within the nebula produces a lot of emission but suffers little absorption, and so more photons emerge to eventually reach our detectors.

There's one further refinement that we can make to the emission coefficient, and that is to include the radiation, which enters a particular volume element and is then scattered so that it leaves the volume element traveling in our direction (the scattering "double-agent game" mentioned above). The scattering coefficient σ can simply be added to the emission coefficient to give us what amounts to a modified emission coefficient, which we can represent by the symbol eta, or "η." The final thing to do is to add this total emission that comes from within the nebula itself to the starlight that has passed through and emerges from the entire nebula (as given by Equation 1) to give us one grand equation:

$$I_{out} = I_{in} \times e^{-\tau} + S \times (1 - e^{-\tau}) \tag{6}$$

Without taking too much notice of the actual details of this equation, what is it actually telling us? If the nebula or indeed any absorbing medium is optically very thick, such as when the value of "τ" is very high, then the value of $e^{-\tau}$ becomes so small that it can be assumed to be equal to zero and then I_{out} is simply equal to "S," the source function. This means in effect that the only radiation that emerges from the medium is that which is emitted by the layer nearest to us; emission from further back is completely absorbed.

By contrast, for a medium that is optically very thin, i.e., τ is very small indeed, $e^{-\tau}$ becomes approximately equal to e^0, which as we saw in chapter *Starlight by Numbers*, is equal to 1. So now I_{out} becomes equal to I_{in}; i.e., all the light that enters the medium gets through. Notice also that the term involving the source function vanishes; this simply means that the medium is so rarefied and thus there are so few gas atoms to ionize that there is essentially no emission from within the medium itself.

So, to sum up; starlight of what we might call "our" wavelength enters a nebula; some of it is absorbed and some of it is scattered, and "our" beam of starlight is thus diminished. Starlight of a shorter wavelength is absorbed, and then some of it is recycled at "our" wavelength; some of this is absorbed and/or scattered, but some survives to intensify "our" beam of starlight. Finally, starlight of "our" wavelength is scattered so that it joins "our" beam of starlight – again, some of this is absorbed and/or scattered and yet again some survives to further brighten "our" beam of starlight.

Clearly the inside of a nebula is a very busy place. Always remember, finally, that all of the quantities that go into and come out of Equation 6 are dependent on the wavelength of the light that we are dealing with. Figure 5 attempts to sum up the kinds of things that can happen within some gaseous material capable of absorbing, scattering, and emitting radiation.

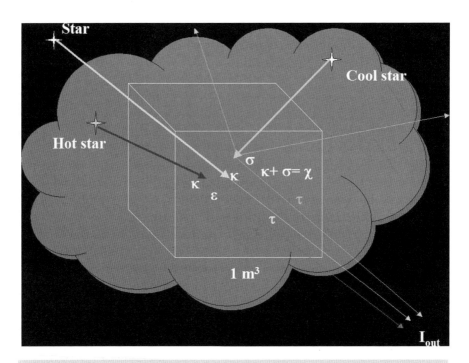

Figure 5. Here we attempt to illustrate the various processes that could take place within one-cubic meter of, for example, a galactic nebula. Light of wavelength "λ" from stars, either within the nebula itself or beyond its far side will suffer absorption and a decrease in intensity according to the value of the absorption coefficient "κ_λ." Some of this radiation will be emitted at some other wavelength and thus contribute to the emission coefficient "ϵ_λ" of the cube. Some radiation will undergo scattering, according to the value of the scattering coefficient "σ_λ," which may contribute to the overall extinction "χ_λ" within the cube, or if scattered in the appropriate direction will further contribute to the emission. All of the radiation that emerges from the near face of the cube will, in turn, suffer further absorption/scattering on its way out of the nebula.

Symbiotic Nebulae – An Illustration of Optical Depth in Action

Symbiotic stars are a group of around 150 fascinating variable stars. They are binaries in which one component is a red giant star – often a semiregular or a Mira-type variable and the other is sometimes a bit of a mystery. What is clear is that this other star is much smaller and also very much hotter, so possible candidates are a hot white dwarf, similar to the central stars of planetary nebulae, and in some instances a neutron star.

The hot component of a symbiotic binary is a source of high energy ultraviolet (UV) and sometimes even X-ray photons, while the cool red giant component, in common with many cool giant stars, emits a slow moving but relatively dense stellar wind. Among various chemical elements and simple molecules this wind will contain significant amounts of neutral hydrogen gas, whose atoms are prime targets for UV photons. These photons ionize the gas atoms and so are absorbed, resulting in a drop in the intensity of ultraviolet radiation as we move away from the hot star.

Electrons in this ionized portion of the cool giant wind or "symbiotic nebula" subsequently recombine with the ionized hydrogen and in doing so return the lost energy. However, this "energy debt" is not returned as a lump sum; we don't get our UV photons back. Instead it comes as "small change" in the form of a series of lower energy visible light photons at discrete wavelengths – emission lines. So part of the cool giant wind is made to glow in visible light, just like the gas that makes up a planetary nebula. However, unlike a planetary nebula, the source of the ionizing radiation is offset from the center of the ionized material, and this provides an excellent yet simple demonstration of optical depth in an absorbing medium at work.

In the 1980s two Canadian astronomers, A. R. Taylor and E. R. Seaquist, developed a simple model for the shape of symbiotic nebulae in order to shed "light" on the fact that quite a few symbiotic stars were known to be radio sources. The origin of this radio emission is itself an interesting example of how starlight is recycled within a nebula. The ultraviolet photons ionize the gas atoms and clearly, for a time before electrons recombine with atoms to give us optical wavelength emission, these electrons will be moving around "free."

However, because of the local presence of many ionized atoms, there will be extensive local electric and magnetic fields. Electrons, which are themselves, of course, negatively charged, change their motion under the action of these electric and magnetic fields; in other words, they accelerate or decelerate, and when this happens they emit electromagnetic radiation, which is of low energy or in fact radio emission. Emissions like this, which result from accelerating and/or decelerating electric charges, are known by the German name bremsstrahlung, or "braking radiation," and because each individual electron involved starts off as a "free" electron and ends up as such, albeit with an altered state of motion, it is also known as "free–free emission."

Now back to the nebula. We know that what determines the optical depth in a medium like this is the linear absorption coefficient, together with the distance traveled through the medium. This, in turn, will determine by how much the initial intensity of a beam of, in this case UV radiation, will drop, and of course a sufficiently high optical depth can cut down a beam of radiation to virtually nothing. The beam simply runs out of photons. If the absorption coefficient is large, then a relatively large optical depth is achieved over a relatively small distance, whereas a smaller absorption coefficient means that the beam travels further before the photons give out.

In a symbiotic system we basically have the kind of situation shown in Fig. 6; UV photons leave the hot star traveling in all possible directions as they head into the cool star's outflowing wind. What is clear and pretty obvious is that as we get closer to the cool giant star, the density of the wind material will increase and, consequently, so will the linear absorption coefficient, whereas further away the wind density drops and so does the absorption coefficient. This means that beams of UV photons leaving the hot star that travel toward the cool giant will peter out after a much shorter distance than those that head away from the cool star. The result is that for a hot star with a relatively low UV luminosity, an elliptical-shaped nebula will result, as shown in Fig. 6a.

The boundary or edge of an ionized nebula is often referred to as the *ionization front,* and in this particular situation its location is determined entirely by where the supply of UV photons gives out. In this case the nebula is said to be *radiation bounded*; the nebula glows in the visible part of the spectrum, but all the UV photons have been absorbed.

Keeping conditions in the cool giant wind the same in terms of composition and density, but this time with a hotter star of higher UV luminosity, it's possible for UV photons traveling directly away from the cool star to actually avoid absorption and escape from the system. In this direction the density of the wind material is dropping, and so, too, is the absorption coefficient; the UV photons simply run out of targets and so this part of the nebula would have no sharply defined ionization front but would simply become fainter as it faded away, as indicated in Fig. 6b. This part of the nebula is then said to be *density bounded*. Finally, for a hot star that can hit the cool giant wind with a very high UV photon luminosity, most of the nebula becomes density bounded, and there remains just a cone-shaped ionization front, produced by the very densest part of the wind and by the shielding effect of the cool giant itself, as shown in Fig. 6c.

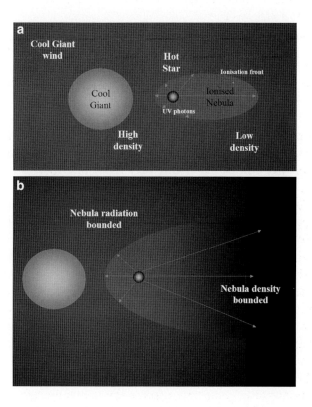

Figure 6. Here we see the effect of varying absorption coefficient and optical depth on the shapes of model ionized nebulae within symbiotic stars. For a given cool giant wind density distribution, the resulting nebula will be either entirely radiation bounded (**a**) partly density bounded (**b**), or almost entirely density bounded (**c**), depending on the intensity of ionizing photons that are emitted by the hot star in the system.

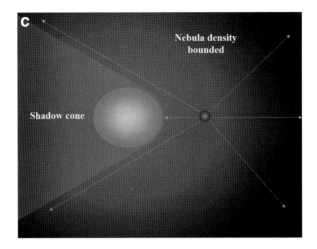

Figure 6. (continued)

This simple model gives a very effective illustration of the kinds of things that can happen in all types of emission nebulae, where densities and chemical composition can vary from one location to another. It also gives us a good start in understanding what happens to our river of starlight between leaving the photosphere and reaching our detectors. Finally, the value of the absorption coefficient and the way that it varies with wavelength is crucial to what happens to radiation, as it makes its way up through the deeper denser layers of a star and out through the photosphere.

Now it's time to take a look *inside* a star.

Key Points

- The degree to which intervening material such as gas or dust is opaque or transparent to starlight depends mostly on wavelength.

- Absorption takes place when photons are removed or lost from a beam of starlight by processes such as ionization.

- The fractional drop in the number of photons in a beam of starlight as it passes through a small unit distance of an absorbing medium is called the linear absorption coefficient κ_λ. A high value implies a relatively opaque medium.

- Provided conditions such as density, chemical composition, and so on within an absorbing medium are constant, then the product of absorption coefficient multiplied by distance into the medium is called the optical depth τ_λ. If the value of τ_λ is greater than 1 the medium is said to be optically thick; otherwise it is optically thin.

- Photons that are not absorbed but which are caused to change direction by atoms or dust grains are said to be scattered.

- The fractional drop in the intensity of a beam over a small unit distance due to scattering processes is called the scattering coefficient σ_λ. This can simply be added to the absorption coefficient to give the extinction coefficient χ_λ.

- Short-wavelength radiation can photoionize gas atoms in an absorbing medium. Recombination of atoms can produce optical wavelength photons, which are observed as extra emission from within the medium.

- The quantity of nebular radiation that emerges depends on the ratio of the emission coefficient to the extinction coefficient, i.e., the source function within the absorbing medium.

- All quantities such as optical depth, source function, and so on are strictly wavelength dependent.

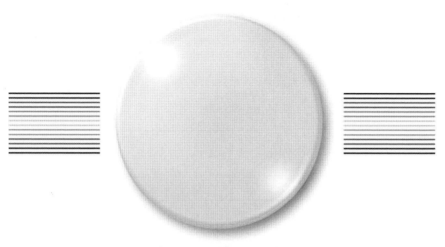

First Look Inside a Star – The Atmosphere

As astronomers, we soon learn that the blindingly brilliant visible face of the Sun is called the *photosphere*, and we could be forgiven for thinking that this is some kind of precisely defined surface, like the surface of a planet. However, all stars including the Sun are entirely gaseous, and the photosphere consists of what amounts to the inner part of a star's surface layers – in effect, its atmosphere. It is a layer of gas of finite depth that is relatively dense, and so it is the principal source of the continuum part of a star's spectrum. However, photons that are emitted within the photosphere and which escape from its topmost layers do so as a result of having traveled along a range of paths of both varying distance and optical depth. This has interesting consequences for the overall appearance of the Sun's photosphere and maybe that of other stars, too.

Both the density and the temperature of the gases of which the photosphere is composed decrease toward its upper regions, and so it is also the main source of the absorption lines in the spectrum. In the case of the Sun and very likely some other stars, too, the atmosphere extends above the photosphere, to layers that are much less dense. These upper layers, the chromosphere and the corona, produce emission-line spectra, and the only way they can do this is by being much hotter than the photosphere.

How can this happen? We've also seen that the continuum part of a star's spectrum resembles that of a perfect blackbody, but there are differences. The atmospheres of distant stars cannot be observed in the

K. Robinson, *Starlight*, Patrick Moore's Practical Astronomy Series,
DOI 10.1007/978-1-4419-0708-0_7, © Springer Science+Business Media, LLC 2009

same kind of detail as that of the Sun, but by taking a closer look at how stellar spectra differ from blackbody spectra, we can discover some of the fascinating processes going on in stellar atmospheres that give rise to these differences.

Simple Radiative Transfer in the Photosphere

Remember in chapter *Space – The Great Radiation Field* we said that unlike the vast majority of stars (where we can only measure the total flux that we receive from them), in the case of the Sun we can measure the specific intensity of radiation from various parts of the solar disk. Most amateur astronomers have, at some time, used a small telescope to safely project an image of the Sun onto a white card or screen in order to observe sunspots. In the absence of specialist equipment that can be used to observe features such as solar prominences, the only other really noticeable feature of the white light solar disk is the fact that the region around the periphery or the limb appears distinctly dimmer than the central regions. This *limb darkening* is, in fact, a relatively simple example of radiative transfer in action.

The gases – mostly the hydrogen and helium that make up the solar photosphere – are not transparent; they absorb radiation that is coming up from the hotter layers down below. What's more, as we descend into the photosphere (what a journey that would be!), the gases become denser, which means that their ability to absorb radiation increases. In other words, with increasing physical depth the linear extinction coefficient "χ" at a given wavelength, for example 5,000 Å, also increases, with a resulting much steeper rise in the optical depth.

So at the topmost layer of the photosphere the optical depth is effectively zero (though not quite as we'll see shortly), and photons can escape freely but go down a certain distance, and the optical depth eventually reaches a value of one. At this depth, about one third of the radiation that starts here manages to make it to the surface, while further down, the photosphere becomes optically thick; by the time we reach an optical depth of 5, less than 1% of the radiation gets to the surface, and at an optical depth of 20, virtually no photons get through. They are instead absorbed and possibly recycled into several emitted lower energy photons or scattered into other directions – possibly back down to deeper layers. This effectively marks the "bottom" of the photosphere, which for the Sun lies at a depth of around 450 km, and it also means that virtually all of the radiation that leaves the Sun's surface comes from this remarkably thin layer of gas. The graph in Fig. 1 shows how the optical depth in the Sun's photosphere increases dramatically with physical depth.

Radiation, which is emitted from some way down in the photosphere and which does make it to the surface may get there by a variety of paths

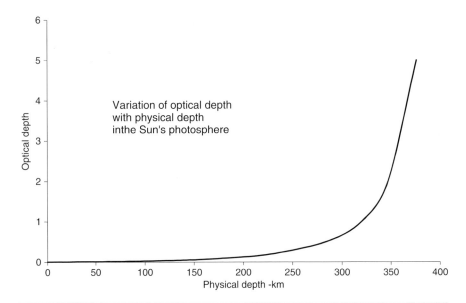

Figure 1. This graph shows the dramatic rise in optical depth, as "we descend" through the photosphere. At a physical depth of around 450 km, the solar material is essentially totally opaque. This means that what we "see" of the Sun in visible light comes solely from the very topmost surface layers; compared to the Sun's size, the photosphere is an incredibly thin layer of gas.

that are of different lengths and thus different optical depths. The basic situation is shown in Fig. 2, where the overall depth of the photosphere compared to the curvature of its topmost layer is greatly exaggerated. We have a small volume – say, a cubic meter situated within the lower part of the photosphere. The two extreme routes by which radiation can leave this volume in order to get to the surface are toward "A" and toward "B."

Clearly, path A marks not only the shortest route of escape but also has the smallest optical depth for the photons, and an observer, who is situated some 93 million miles away in this direction will observe these photons apparently coming from the center of the Sun's disk. By contrast route "B" takes the photons through the lowest densest region of the photosphere and is thus the route of greatest optical depth. Very few, if any, photons will escape along this route, which to our distant observer will now appear to come from the limb of the solar disk. Between these two extremes the specific intensity of radiation and thus the apparent brightness of the Sun's disk decrease from the center to the limb. The resulting limb darkening is, however, not as extreme as it might at first seem, because within the photosphere itself there is a lot of scattering going on, and closer to the surface of the photosphere, there will be radiation that gets scattered in B's direction.

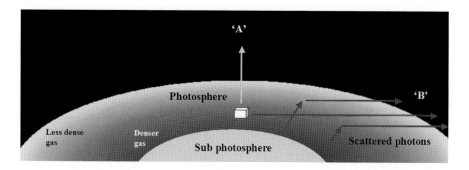

Figure 2. Paths "A" and "B" mark the routes to the surface of smallest and largest optical depth, respectively, for photons leaving the small unit volume. Very little radiation will survive the journey along route B, resulting in the Sun's limb appearing darker than the center of the disk. In the outer layers of the photosphere, photons will get scattered into B's direction, making the limb appear not quite so dark as it otherwise would.

With what we learned in the previous chapter, we can in fact be a bit more specific about the effect of limb darkening. To do this we take advantage of the fact that the actual known depth of the photosphere is around 450 km, which makes for a very physically thin layer when compared to the Sun's radius of around 700,000 km. This enables us to make use of what the professionals call the *plane-parallel approximation*, which means that we can essentially go back to the slab-shaped absorbing medium we used in chapter *The Photons Must Get Through – Radiative Transfer* and perhaps what's more to the point, the mathematics becomes very much simpler.

Figure 3 shows the layout. Notice that the trade off for using this simplification is that light paths, which are parallel to the plane of the photosphere, go off to infinity, without ever reaching some kind of surface. So the plane-parallel approximation is okay provided we don't make the angle "θ" (theta) in Fig. 3 too close to 90°.

The advantage of the plane-parallel approximation is that we can assume that the topmost layer of the photosphere is flat. This makes for

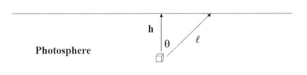

Figure 3. Because the Sun's photosphere is physically relatively very thin, it can be represented with the "plane-parallel approximation," which is shown here. We have, as before, our one meter cube situated at a vertical depth "*h*" below the surface. Radiation travels along the path of length "*ℓ*," which makes an angle θ to the vertical direction.

simple trigonometry because any path length through the photosphere "*ℓ*," is given simply by

$$\ell = h/\cos(\theta) \tag{1}$$

The linear extinction coefficient will vary with depth in the photosphere, but if we assume that it varies in the same way all over the Sun, then we can in the current situation simply multiply its average value χ by the path length ℓ to get the optical depth τ_ℓ.

$$\tau_\ell = \chi \times \ell = \chi \times h/\cos(\theta) \tag{2}$$

But $\chi \times h$ is the optical depth straight up through the photosphere, which we can call τ_0. Thus

$$\tau_\ell = \tau_0/\cos(\theta) \tag{3}$$

So τ_ℓ of course will always be bigger than τ_0 for a given physical depth in the photosphere. The flip side to this is that in order to go down to where the photosphere has some given optical depth – say "1" – you need to go to a greater physical depth, if your route is in the direction that is normal to the plane of the photosphere, as shown in Fig. 4. This means that along this direction, we are seeing deeper into the Sun, and these deeper layers are correspondingly both denser and hotter. They thus appear brighter. A careful study of limb darkening can actually enable astronomers to determine temperatures within the photosphere.

Unfortunately limb darkening cannot be observed for all but a few distant stars. These "few" are eclipsing binary stars, where the effect of limb darkening on one or both components manifests itself in the shape of the system's light curve. A computer can be used to model the light curve of an eclipsing binary system, based on physical parameters such as mass, radius, separation of component stars, and also limb darkening. The most famous and widely used eclipsing binary light curve modeling program is that which was developed over many years by R. E. Wilson and E. J. Devinney. This program is freely available on the Internet.

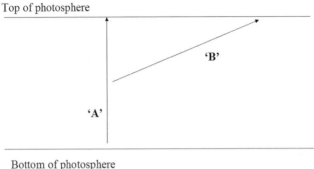

Top of photosphere

'B'

'A'

Bottom of photosphere

Figure 4. Paths A and B have the same optical depth, but as can be seen, path A enables us to see down to a greater physical depth, where the solar material is hotter, denser, and brighter. The result is that the center of the Sun's disk is the brightest part.

Because the photosphere is a layer of gas of finite thickness, its temperature varies from around 4,500 K at the topmost layer to around 9,000 K at the bottom, and the average photospheric temperature is that fairly familiar figure of 5,800 K. An important point to note here is that because the formation of absorption lines due to various chemical elements is very temperature dependent, it means that different lines are produced in different layers of the photosphere. An old term – the *reversing layer*, which suggests that all absorption lines are produced within one single layer – is not used these days, though you may still come across it in the literature. The photosphere itself in fact forms the lowest of the three layers of the Sun's atmosphere; above the photosphere, strange things happen.

The Chromosphere and the Corona

If the temperature continued to fall above the photosphere, then all that these cool lofty gases would do is to strengthen slightly the absorption lines that form in the photosphere itself. The effect would be small, too, because the optical depth from the uppermost layer of the photosphere all the way to the top of the Sun's atmosphere is only of the order of 0.005; so we're dealing with stuff that is pretty optically thin.

However, both the middle and upper parts of the solar atmosphere are emission-line regions, and this means that atoms within this material are being ionized. In what we might call a more "familiar" kind of emission-line region, such as a planetary nebula, photoionization by high-energy ultraviolet photons does the job. However, it was realized many years ago that the UV flux values from stars such as the Sun are insufficient to make photoionization a significant "player" here. The other way to ionize a gas is to raise its temperature just as Bunsen and Kirchoff did by vaporizing chemical salts in a hot flame. This process is called *thermal ionization*, and clearly some process must be pumping energy into the Sun's upper atmosphere to raise its temperature to at least 10,000 K. Above this temperature, hydrogen atoms start to ionize.

The layer above the photosphere is seen famously as a thin red arc just before and just after totality in a solar eclipse. This is the *chromosphere*, which extends for about 10,000 km above the photosphere. The strong red color of the chromosphere, caused by our old friend the Hα line at 6,563 Å, suggests that the temperature is of the order of 10,000 K, and indeed temperatures toward the top levels of the chromosphere are estimated to reach around 20,000 K. Above the chromosphere, the so-called *transition region* eventually gives way to the Sun's *corona*, which extends outward for several million kilometers. As we head toward the "roof" of the solar atmosphere, temperatures themselves go through the roof to the order of several million degrees. This is known because the highly rarefied gases of the corona produce emission lines in the far ultraviolet and even in the X-ray regions of the e-m spectrum. These lines are produced by heavier elements whose atoms have lost several electrons, and such a high degree of ionization can only result here from exceedingly high temperatures.

What causes the chromosphere and the corona to have such high temperatures is still the subject of a great deal of debate. In the late 1940s a perhaps rather unusual source of at least chromospheric, if not coronal, heating was suggested; this was acoustic waves – in other words, "sound waves" coming from the photosphere.

Observations of the photosphere under very good seeing conditions reveal a speckled appearance, called the "solar granulation." Each "speckle" is the top of a "blob" (professional astronomers do indeed use the word "blob" in this context) of gas – also known as a convective cell – which has risen up through the photosphere as a result of heating in a lower layer. This process generates a great deal of turbulence within the solar gases, which in turn generates acoustic waves. At the top of the photosphere, the falling density amplifies these acoustic waves as they travel upward, and their energy heats the chromosphere as it dissipates.

How does an astronomer set out to detect the presence of sound waves traveling up through the chromosphere, short of dangling a microphone into the region and listening in?

In a transverse wave, such as the rope wave that we discussed in detail in chapter *From Light to Starlight*, the motion of any point on the wave is at right angles to the direction in which the wave is traveling. For acoustic waves that travel through a gas, it is the gas atoms that move as the wave moves along. In this case, though, each gas atom moves backward and forward in the same direction in which the wave travels; as a result, an acoustic wave is an example of a *longitudinal wave*. Instead of the crests and troughs, which are the key features of a transverse wave, a longitudinal wave consists of a regularly spaced series of alternating zones, where the gas is denser than normal and more rarefied than normal, as shown in Fig. 5. These zones are respectively called *compressions* and *rarefactions*, and the distance between the centers of two neighboring compressions determines the wavelength of the acoustic wave, though with acoustic waves it is much more common to speak of the frequency, which just as with transverse waves is equal to the wave's velocity divided by the wavelength.

Now imagine an acoustic wave traveling up through the chromosphere, which itself is emitting, for example, Hα emission-line radiation. This ascending acoustic wave will produce within the chromosphere a series of compressions and rarefactions; in the compressions, the gas atoms over a limited region will temporarily squeeze together, and the resulting slightly higher density will slightly increase the brightness and width of the emission line. By contrast when a rarefaction passes, the atoms spread out slightly, lowering the local density and producing a slight narrowing and drop in the brightness of the Hα line. So, in theory, careful observation of emission lines in the chromosphere should reveal regular variations in their brightness.

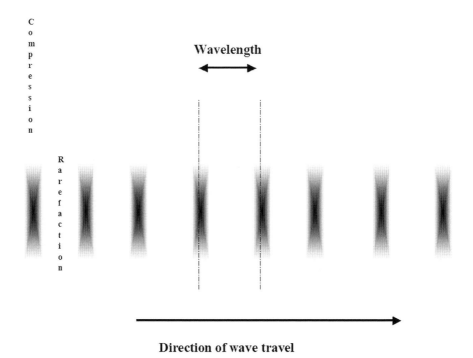

Figure 5. In a longitudinal wave such as a sound wave, the gas atoms oscillate backward and forward along the line in which the wave is traveling. The result is a moving sequence of zones, where the gas is denser than average (these are called *compressions*), alternating with zones of lower than average density gas (these are called *rarefactions*).

This process, which is illustrated in Fig. 6, is a very big simplification of what are in fact very difficult and complex observations, which also require very detailed analysis. However, such observations have been carried out, and they do indeed suggest that acoustic waves are at least partly responsible for heating the solar chromosphere. Many argue, though, that acoustic wave heating is not in itself sufficient and that the Sun's magnetic field must in fact play a more dominant role, particularly with regard to heating the gases in the corona. The acoustic wave camp in turn argues that for smaller stars, which rotate relatively rapidly and in which convection itself plays a much less significant role, magnetic fields may indeed be the dominant heating mechanism. However, for large slower rotating stars such as red giants, acoustic wave heating is more likely to be the key to chromospheric heating. This debate will probably go on for some time yet.

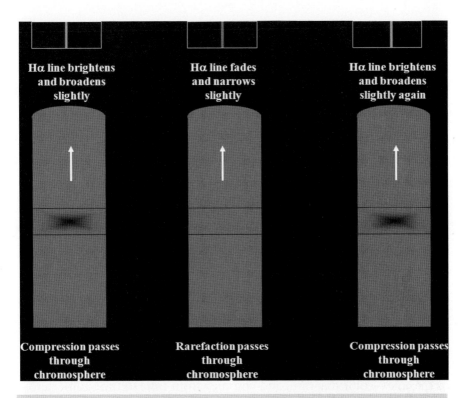

Figure 6. Acoustic waves, i.e., compressions and rarefactions passing up through the chromospheres, cause slight variations in the intensity and width of, for example, the Hα line. In reality, the situation is much more complicated than what is depicted here, and careful observations followed up with detailed analysis is needed in order to unequivocally identify these variations.

Time now for a bit of a gear change. With distant Suns, we don't have the luxury of being able to study their atmospheres in anything like the kind of detail provided for us by our own Sun. Yet by making use of the simple kinds of observations described in chapter *A Multitude of Magnitudes for the Colors of Starlight*, together with the application of a bit of physics, we can, as we'll now see, discover some fascinating things about what goes on in the photospheres of other stars.

Yet Another Kind of HR Diagram

In chapter *A Multitude of Magnitudes for the Colors of Starlight* we didn't say too much about the U − B color index. It's clearly a measure of how much radiation is emitted by a star in the near ultraviolet part of the spectrum, so its value would be negative for those stars that emit more radiation at these wavelengths. As approximate blackbodies, we would expect these stars to be very hot, and as a result we'd also expect them to be very luminous stars. So in addition to seemingly being yet another measure of temperature (we'll see shortly, however, why this is in fact not the case), the U − B color index should also serve as a measure of the luminosity of a star − another of these very important stellar parameters.

So if we were to plot U − B against B − V for a sample of stars, we would in effect be plotting the luminosity of each star against its temperature. As we've seen, luminosity is related to a star's absolute magnitude, and as B − V is related to stellar temperatures, it is also related to the stellar spectral classes. So we are essentially plotting a kind of HR diagram which, because it involves plotting one color index against another, is referred to as a *color–color diagram*. A plot of U − B vs. B − V is sometimes referred to as *the* color–color diagram, and you may from time to time come across other color–color diagrams involving other color indices (not necessarily in the visible spectrum), which are used by astronomers to investigate various regions of the continuum part of a star's spectrum.

The first thing to do is to plot on the same diagram the U − B vs. B − V colors for perfect blackbodies, together with those for main sequence stars. This we have done in Fig. 7. The first thing to note is that the plot for blackbody spectra is a smooth gentle curve − indeed, it's not too different from a simple straight line. The second thing is that the plot for real stars generally falls below that for blackbodies. Some authors may adjust the magnitude scales for real stars so that for example the color indices of stars such as the Sun match those of a corresponding blackbody. This then has the effect of raising parts of the color–color plot for real stars above that for blackbodies.

Strictly speaking, however, the real star color–color plot will always fall below the blackbody plot. What this is telling us is that the luminosity of most stars as measured here by the value of U − B is lower than it would be if the stars were perfect blackbodies. This is to be expected, because as we can by now well appreciate that the non-blackbody nature of real stellar spectra results from the simple fact that stars as emitters of radiation are not as efficient as blackbodies. Aside from these two things, the plot for the main sequence stars looks a bit weird. With the help of a bit of physics, though, this odd shaped U – B vs. B – V curve will teach us a great deal about what's going on within the photospheres of stars.

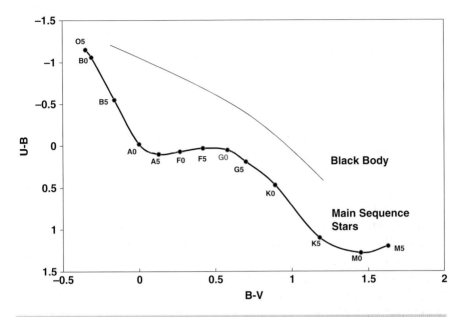

Figure 7. The color–color diagram from main sequence stars and for blackbodies.

A Photosphere Full of Hydrogen Atoms

Among the first things to be identified in the spectra of stars were absorption lines due to hydrogen. This means that the photosphere, or to be more specific, its upper layers, must contain hydrogen gas, which is both cooler and of lower density. It was also soon realized that the intensity of the hydrogen absorption lines varied as one moved through the spectral sequence; they start off by being fairly weak for the very hottest class O and B stars and increase to maximum intensity at class A0. They then progressively weaken as we move to cooler stars. What this tells us is that the ability of hydrogen gas or, in other words, hydrogen atoms to absorb radiation is very dependent on the temperature of the hydrogen gas itself.

A hydrogen atom like all atoms consists of a tiny nucleus, which contains almost all of the atom's mass and carries a positive electric charge. Way out on the periphery of our hydrogen atom, there "lies" an electron, a tiny particle of very little mass but with a negative electric charge equal in magnitude to that of the positive charge in the nucleus. So generally speaking our hydrogen atom is an electrically *neutral* atom. We saw in chapter *From Light to Starlight* when we described the photo-electric effect that it is possible for light (i.e., photons) with sufficient energy to actually remove electrons from atoms. In general the process of removing electrons from atoms is called *ionization* (the end product is called an *ionized atom* or simply an *ion*) and if it is light that does the ionizing, the process is, as we saw in the previous chapter, called *photoionization*. This process, however, does not produce hydrogen absorption lines.

The Hydrogen Absorption Lines

One of the most important discoveries to come out of the great experimental work of Bunsen and Kirchoff was that absorption lines are not distributed randomly across a spectrum. For any given chemical element, and this includes hydrogen, of course, the individual lines have fixed wavelengths that do not change. In 1885 the Swiss physicist and mathematician Johan Jakob Balmer was able to derive a formula that successfully calculated the wavelengths of the absorption lines in the visible spectrum of hydrogen; as a "reward" these lines are now known as the *Balmer lines*.

Balmer's formula was what is known in the trade as an *empirical formula*, in that it involved the use of numbers that, at the time, really could be regarded as "fiddle factors," because their physical significance was not understood. Balmer's formula worked, though, and that was a step in the right direction. In that same year of 1885, the Danish physicist Niels Bohr was born, who some thirty or so years later gave physical meaning to those empirical numbers in Balmer's equation.

The fact that the Balmer lines (and indeed all absorption lines) have wavelengths that are fixed means that hydrogen atoms like those of all other chemical elements are only absorbing light of certain specific wavelengths. In other words, to produce absorption lines in a spectrum, atoms can only absorb photons of very specific energies; they do not absorb photons of energy "somewhere in between" these specific energies. We know that a photon with sufficient energy will ionize the atom by removing an electron, and this, in fact, suggests very strongly that the whole business of atoms absorbing photons involves the electrons and not the nucleus. So if our hydrogen atom absorbs a photon of a specific energy (thus contributing to one of the Balmer lines), the electron must have gone from a state of lower energy to a state of higher energy.

The difference between the "before" and "after" states is equal to the energy of the absorbed photon, but because the absorbed photons have to be of very specific energies, it would seem very likely that the "before" and "after" energy states for the electron itself must also have very specific values. In other words, the electron in our hydrogen atom (and this also applies to any electron in any atom) can only reside in very specific energy states within the atom (energy states are called *energy levels*). Niels Bohr followed a more theoretical and of course a more rigorous line of reasoning to reach the conclusion that electrons can indeed only exist within atoms in certain specific energy levels. The exact nature of these energy levels, including the energy that an electron has when it resides in one, is determined by the laws of quantum mechanics. This can become pretty complicated for elements other than hydrogen, which involve multi-electron atoms. For hydrogen, with its one single electron, though, things are mercifully much simpler.

The Ups and Downs of a Hydrogen Atom

Every hydrogen atom has a set of discrete energy levels, in any one of which its single electron can reside. Which energy level is occupied depends on the energy of the electron itself, and to further facilitate things, the energy levels are each given a number in order to identify them. This number "n" is called the *principal quantum number*; $n = 1$ identifies the level in which the electron would have the lowest possible energy within the atom. $n = 2$ corresponds to the next higher energy level and so on.

A very important feature of the energy levels is that they are not equally spaced in terms of the actual energy differences between adjacent levels. The lower levels are relatively widely spaced in terms of these energy differences; they then become increasingly more closely spaced for the higher energy levels and finally merge together at the very limit of where the electron can be said to be actually attached to the atom's nucleus. A higher energy than this means that the electron has in fact "escaped" and so the atom has been ionized. The traditional way to represent the energy levels in an atom is as a series of horizontal lines (rather like shelves) on a diagram, where the bottom line represents the nucleus of the atom and the top of the diagram represents the "outside world" away from the atom. (This outside world is often referred to by physicists as the "continuum," and it should not be confused with the continuum part of a spectrum.)

Making an absorption line is now quite straightforward; our hydrogen atom sits there with its electron in one of the energy levels, and then along comes a photon of just the right energy (wavelength), which gets taken out (i.e., absorbed by "moving" the electron to a higher energy level). Multiply this event many times within a vast population of hydrogen atoms and a radiation field containing many photons of the right wavelength, and lo and behold we have an absorption line. Which absorption line we get depends, of course, on the two energy levels that are involved in this particular *electron transition*. The Balmer lines all result from the electron starting off in the $n = 2$ level; the transition from $n = 2$ to $n = 3$ gives us the well-known Hα line at 6,563 Å; that from $n = 2$ to $n = 4$ produces the Hβ line at 4,861 Å; and so on.

Electron transitions that start from other levels give rise to other series of hydrogen absorption lines. For example, transitions starting on the bottom or $n = 1$ level produce the Lyman lines, which are visible in the ultraviolet part of the spectrum, and those starting on the $n = 3$ level produce the Paschen lines in the infrared. The relative layout of the hydrogen energy levels and the main electron transitions are summed up in Fig. 8.

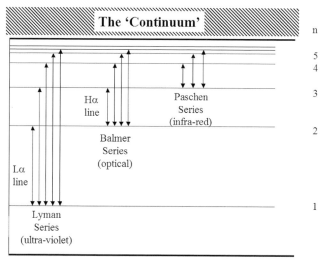

The Nucleus

Figure 8. This is how the electron energy levels in an atom are represented. (Such "Grotrian" diagrams, as they are known, for multi-electron atoms are much more complicated than that for hydrogen, which is shown here.) Notice that the energy levels crowd ever more closely together as we get nearer to the continuum. The familiar Hα line transition at 6,563 Å together with that for the Lyman alpha or Lα line that was mentioned in chapter *From Light to Starlight* are indicated. An *upward arrow* indicates absorption of a photon, with a downward arrow indicating corresponding photon emission.

Finally, as they say, "What goes up must come down." An electron that has been elevated to a higher energy level will only stay there for a very brief interval – something of the order of a one hundred millionth of a second. It will then drop back down to a lower level, and in dropping it emits a photon, quite possibly one of exactly the same wavelength as that which was absorbed. For stars, however, there is a big difference between the original absorbed photon and the subsequent emitted one. The vast majority of the absorbed photons come from the deeper, hotter layers of the photosphere, so they come predominantly from one direction. Had they not been absorbed at all, they would continue on their journey outward to become part of the star's continuous spectrum. The re-emitted photons, on the other hand, are highly unlikely to head outward in the original photons' direction, and indeed they have a roughly 50% chance of heading back down into the star itself. The net result is a deficit of photons that are headed our way, and so we "see" an absorption line.

The Ups and Downs of a
Hydrogen Atom in a Photosphere

Now our hydrogen atom, together with all its buddies, actually finds itself within the photosphere of a star. If the star is a relatively cool one, say, of spectral class K or M, then the electrons for the most part will be in the $n = 1$ – the lowest energy level. This is so because this is the "normal" state for hydrogen atoms; to get the electrons into the higher energy levels takes extra energy, maybe in the form of photons coming in from outside the atom or by raising the temperature. This means that in relatively cool stars, there won't be many hydrogen atoms with their electron in the $n = 2$ level, which is the "starting point" for the formation of the Balmer lines, and so the Balmer lines in the spectra of these stars are relatively weak.

Now let's raise the temperature of the photosphere; this means that our hydrogen atoms together with atoms of other elements, too, are moving around faster. "Collisions" between atoms will take place that, provided the photosphere temperature is not too high, will not be energetic enough to thermally ionize our hydrogen atoms, but they will carry sufficient energy to "lift" those electrons from the $n = 1$ level to the $n = 2$ level. This process is called *thermal excitation* or *collisional excitation*. By the time the temperature reaches 10,000 K the $n = 2$ energy levels of our vast population of hydrogen atoms are themselves maximally populated. These atoms are now primed and ready to absorb incoming Balmer line wavelength photons (one should probably use here the expression "upcoming," because most of these photons are coming from the deeper, hotter layers of the photosphere).

Stars of spectral class A0 have a photospheric temperature of around 10,000 K, and these stars do indeed have the strongest, darkest Balmer lines in their spectra. With increasing photospheric temperatures, increasing thermal excitation makes the electrons migrate further up the energy levels, and so the Balmer lines weaken for spectral classes B and O. For the very hottest photospheres most of the hydrogen atoms are ionized, and so only weak absorption lines can form.

A hydrogen atom whose electron is in the $n = 2$ level will be ionized by a photon whose wavelength is either equal to or shorter than about 3,647 Å. This wavelength is in the near ultraviolet part of the e-m spectrum, which means that the radiation field within the photospheres of relatively cool stars won't contain so many of these kinds of photons. Very hot stars will produce large numbers of these photons, but their photospheres don't contain so many hydrogen atoms with electrons in the

$n = 2$ level. With spectral class A stars, though, conditions are just right. These are relatively hot stars, so there are plenty of the right kind of photons, and as we've seen their photospheres contain large supplies of level $n = 2$ hydrogen atoms. This situation has the potential to take out not just individual wavelengths in the visible spectrum but a whole swathe of the near ultraviolet continuum.

There's a peculiar thing about quantum mechanics (just one of many peculiar things, in fact), which is a bit like a resonance effect, whereby, for example, a bridge can be made to sway or vibrate as a result of people walking across it in step and with just the right rhythm. A population of level $n = 2$ hydrogen atoms will absorb photons of the critical wavelength 3,647 Å in vast numbers. As we move away from this value to shorter wavelengths, the number of absorbed photons falls off and eventually peters out. The effect of this absorption on the spectrum is to "cut out" a saw tooth-shaped depression in the near ultraviolet continuum. At 3,647 Å the continuum drops abruptly to produce what is known as the *Balmer discontinuity* or sometimes simply the *Balmer jump*; the continuum then recovers as we move further into the ultraviolet. As we've seen, conditions within the photospheres of stars, which are both cooler and hotter than class A0 stars, have fewer $n = 2$ level hydrogen atoms in their photospheres, which makes them less suitable for the Balmer jump to form, so this feature becomes less prominent.

On the other hand, stars hotter than class A0 will have more hydrogen atoms with the electron in the $n = 3$ level, and this can give rise to a "Paschen jump" at 8,212 Å in the infrared, which itself affects the visible part of a star's spectrum. These two jumps are shown in stylized form in Fig. 9. So the jumps or discontinuities that are associated with series of spectral lines can affect a whole region of a star's continuous spectrum, making it differ, perhaps significantly, from that of a blackbody spectrum. Indeed the presence of other chemical elements in the outer layers of stars can give rise to an assortment of jumps that, in turn, depending on conditions of temperature and the relative abundance of said elements can also affect stars' continuous spectra.

The presence of the Balmer jump in stellar spectra can immediately sort out a problem we encountered in chapter *A Multitude of Magnitudes for the Colors of Starlight*.

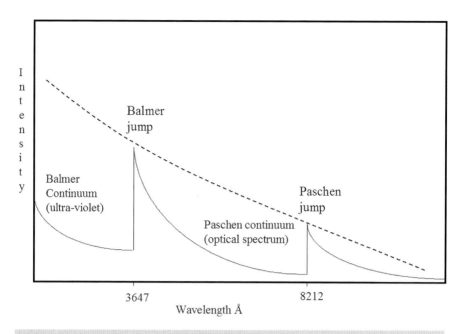

Figure 9. This shows in stylized form the effect of the Balmer jump on the near ultraviolet continuum of a star's spectrum and the corresponding effect on the optical continuum of the Paschen jump. The *dotted line* represents the corresponding spectrum of a blackbody.

The Temperature of Vega

When we tried to estimate the temperature of Vega by using Wien's law, our result fell a long way short of Vega's accepted effective temperature of 10,000 K. Let's actually try plugging this temperature into the Wien's law equation to see what the wavelength of maximum emission for Vega should be. For an effective temperature of 10,000 K this wavelength of λ_{max} in angstroms is given by

$$\lambda_{max} = 2.8973 \times 10^7/10,000 \tag{4}$$

This is equal to 2,897.3 Å, which puts it well within the Balmer discontinuity zone. The observed spectrum of Vega appears to peak at around 4,200 Å; remember, though, Vega is a spectral class A0 star, which means that its spectrum contains the strongest Balmer lines. These lines crowd together as we approach the Balmer jump wavelength of 3,647 Å, and this is a consequence of the merging together of the higher energy levels in the hydrogen atom. These merging Balmer lines cause what should have been a still increasing continuum level to fall away instead, and then the Balmer jump itself finishes the job. So the whole top of the continuum part of Vega's spectrum has been taken out by virtue of Vega's photosphere and those of stars like it, having just the right conditions of temperature in the presence of the right kind of radiation field. This is shown dramatically in Fig. 10.

There is a final footnote here, too; the presence of the Balmer jump means that for many stars the "U" magnitude is not actually representative of what we would expect. More specifically the temperature of these stars is not reflected in the emitted flux values over the "U" wavelength region, and so the U – B color index itself cannot be used as an indicator of a star's temperature.

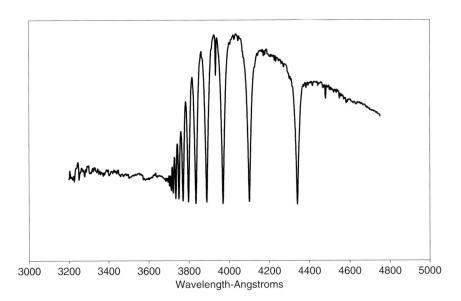

3000   3200   3400   3600   3800   4000   4200   4400   4600   4800   5000

Wavelength-Angstroms

Figure 10. The spectrum of Vega in the blue and near ultraviolet region. This dramatically shows how what would otherwise be a much higher emission maximum at a shorter wavelength is in effect eroded by increasingly closely spaced Balmer absorption lines and finally by the Balmer jump itself, to leave a very much reduced near ultraviolet continuum. (Spectrum reproduced from the STELIB library of the Laboratoiire Astrophysique de Toulouse by kind permission of Jean François LeBorgne.).

Return to the Color–Color Diagram

Another thing that we mentioned in chapter *A Multitude of Magnitudes for the Colors of Starlight* was the choice that Johnson and Morgan made for the wavelength regions covered by the standard U and B filters in their photometric system. These two filters in fact cover wavelength regions that sit on either side of the Balmer discontinuity. The consequence of this is that whereas the B magnitude (and indeed the V magnitude too) of a star generally changes quite smoothly with the star's effective temperature, the U magnitude varies in a more complicated way because of the effect of the Balmer jump. This in fact can help us make sense of the weird U – B vs. B – V plot in Fig. 7.

For the very hottest stars, there is virtually no Balmer jump, because most of the hydrogen atoms in the photosphere are ionized, and the U magnitude for these stars "shines at full strength," giving a U – B value very close to that for a blackbody. As we head toward spectral class A, though, the increasing Balmer jump weakens the U magnitude by much more than would result from a simple fall in temperature. The result is that the value of U – B becomes less negative more rapidly than it does for blackbodies, and so the plot for the stars falls way below the blackbody plot.

At spectral class A0 the Balmer jump has its maximum effect on the U magnitude, but as a result of a dropping photosphere temperature, which itself weakens the "U" flux, the plot continues to fall for a while and eventually "bottoms out" at around spectral class A5. Now the diminishing effect of the Balmer jump enables the "U" region of the continuum to recover and so the U – B color index heads back toward the blackbody value. Beyond spectral class F5 the falling photospheric temperature means that the strength of the "U" region of the continuum continues to weaken, and so this recovery cannot be maintained.

There are a couple of other processes, too, which begin to have their effect as we move to cooler stars. One of these comes from the "trace" of chemical elements other than hydrogen and helium that exist within stellar photospheres. One point where astronomers differ from chemists is that they refer to all elements other than hydrogen and helium (and often to all elements other than simply hydrogen itself) as metals. As we move through the periodic table of the elements, atoms of these metals incorporate increasing numbers of electrons, which are themselves distributed (according to the rules of quantum mechanics) among a range of lower and higher energy levels. These multi-electron atoms are capable of producing various series of absorption lines, depending in part on the abundance of a particular element within a star's photosphere but more so, on the prevailing temperature.

At the high temperature end of the spectral sequence, atoms of metals are multiple-ionized; in other words, they have had the outer electrons removed by both ultraviolet photons and by thermal ionization. The remaining electrons, for the most part, reside in the lower energy levels, which require photons of high energy just to excite them to higher levels. This puts them way out of reach of the lower energy photons of the visible part of the spectrum, and so for the hottest stars, this appears as a relatively smooth, clean continuum save for weak Balmer lines and also lines due to helium.

By contrast, the ultraviolet spectra of very hot stars contain many more absorption lines. With decreasing temperatures in the photosphere the metal atoms progressively "regain" their outer electrons, which can be made to undergo electron transitions by absorbing both visible light photons and photons in the near ultraviolet (i.e., the "U" band), too. So this increasing number of absorption lines due to metals helps to keep the U – B color index depressed, relative to that for a blackbody.

Finally, at the cool, i.e., spectral, class M end of the sequence, temperatures are low enough for simple molecules and "part molecules" (what chemists call "radicals") to exist. These entities produce vast numbers of sets of absorption lines that crowd together to form bands that can "conspire" to knock out whole chunks of the continuum (including the near ultraviolet), an effect known as "line blanketing."

The other effect involves our old friend the hydrogen atom again. As if just to make life interesting, our hydrogen atom is capable of attracting and barely holding onto an extra electron. This "intruder" can be thought of as occupying a very high energy level within the atom, so that very little energy is required to remove it. It thus follows that this so-called *negative hydrogen ion* (written in the literature as "H^-," the minus sign indicating the presence of an extra negatively charged electron) won't survive for very long within the photosphere of a hot star, but it can positively thrive in the photospheres of cooler stars. Because it doesn't take photons of much energy to remove the extra electron (the term used in this case is photo dissociation rather than photoionization), its most significant effect is to absorb photons in the infrared part of the spectrum. However, given a high enough population density of negative hydrogen ions, their absorbing capabilities can easily extend across the visible and into the near ultraviolet, thus further weakening the already low "U" flux levels for the cooler stars. The overall result is that the value of the $U - B$ color index is kept below what it might otherwise be, even for the cooler stars, though the effect is not as dramatic as that produced by the Balmer jump at the high temperature end of the spectral sequence.

Model Atmospheres for Stars

We have seen that a lot can happen in the atmosphere of a star, and much of it causes the overall spectrum of a star to differ from that of a perfect blackbody. The ultimate goal of the stellar astronomer is to produce a model atmosphere for a star, into which various parameters such as temperature and absorption coefficient/opacity for a whole range of wavelengths together with the chemical composition of the star can be entered. The idea then is that the spectrum that would be produced by such a model atmosphere can be made to match as closely as possible the observed spectrum of a star and hopefully, as a result yield valuable information about the star itself.

This is not an easy task. One technique is to effectively represent the photosphere as a series of concentric shells, whose effective temperatures and densities increase with depth. This means, for example, that different absorption lines can be better modeled by having the appropriate absorption coefficient vary with depth in the photosphere. The most serious (and brave) amateur spectroscopists may wish to venture down this road, in which case they should consult more advanced texts. Good luck!

Key Points

- The photosphere of a star is the principal source of both the continuum and the absorption lines in the star's spectrum.

- The Sun's photosphere extends to a physical depth of about 450 km.

- The effect of the photosphere's optical depth means that we can see to a greater physical depth at the center of the Sun's disk than in the limb area; this results in the well-known limb darkening effect.

- The Sun's chromosphere comprises the middle part of the Sun's atmosphere; it is both hotter and less dense than the photosphere and produces an emission-line spectrum.

- The prevailing (though not conclusive) theory is that the chromosphere is heated by acoustic (sound) waves traveling upward as a result of convection processes in the photosphere.

- The color–color diagram plots the U – B color index against the B – V index for stars. It provides a vivid illustration of how stars differ from blackbodies when plotted alongside the blackbody color–color diagram.

- One of the principal causes of continuum absorption in modestly hot stars is the ionization of hydrogen atoms by higher energy photons. This results in a saw tooth-shaped region of the near ultraviolet continuum being removed. This is called the Balmer jump.

- Other processes, such as jumps from heavier elements, the negative hydrogen ion, together with increasing numbers of absorption lines and bands for cooler stars serve to further "degrade" a star's spectrum from that of a blackbody.

- Professional astronomers develop and use "model atmospheres" as a means of matching model stellar spectra to real spectra in order to gain information on the structure of and conditions within a star's atmosphere.

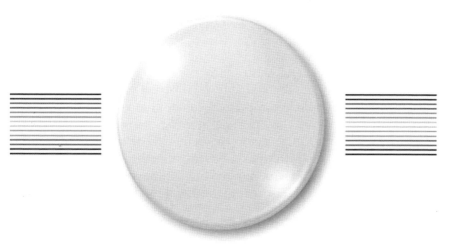

Deep Inside a Star

The great majority of astronomers, both amateur and professional, would probably agree that the most interesting part of the Hertzsprung–Russell diagram is the bit that lies above the main sequence. Here we find most of the intrinsically variable single stars, and what's more, the light variations of these stars are largely due to things that are going on beneath their surface layers. In some cases, for example, with Cepheid variable stars, the process is now fairly well understood, but with the red giant variables, such as Mira-type stars and semi-regular variables, the situation is much less clear. Professionals are using supercomputers to model the kinds of things that go on in the vast envelopes of these stars, but here, perhaps, more than anywhere else, the role of amateur variable-star observers, who make regular observations of these stars, is pivotal in furthering our understanding of what goes on in the upper right-hand corner of the HR diagram.

By contrast, most stars, particularly those that lie along the main sequence of the HR diagram, are not only stable but *very* stable for very long periods of time. For example, geological evidence suggests that the Sun has been shining in more or less the same way for the past 4 billion years or so. Stable main-sequence stars represent the stellar astronomer's level playing field, and clearly it pays to know what it is that makes a star stable in order to better appreciate what can make it unstable. The chief problem here, though, is that unlike the photosphere, there is no way to directly observe the interior of a star.

Recent work involving certain analytical techniques used by seismologists has enabled astronomers to detect oscillations in the surface layers of

K. Robinson, *Starlight*, Patrick Moore's Practical Astronomy Series,
DOI 10.1007/978-1-4419-0708-0_8, © Springer Science+Business Media, LLC 2009

stars; these oscillations are then used to investigate the internal structure of stars.

This area of research, which is called asteroseismology, clearly holds great potential, but generally, astronomers have to use the laws of physics together with what data are available to construct theoretical models of stars, whose "observational properties" can be compared to those of real stars. Central to this are the *equations of stellar structure*, which we shall explain here without going into the mathematical details of the equations themselves. In fact, the whole business regarding the internal structure of stars has quite a lot to do with physics, but as always a little bit of physics brings a big payoff in understanding the astrophysics.

Hot Gas – The Basics

Stars are vast globes of hot gas – indeed, even the very coolest stars are far too hot for any part of them to exist in liquid or solid form. Because of these high temperatures, most of the material that makes up a star is in the form of atoms, which are either totally ionized (they have lost all their electrons) or at least very highly ionized. This mix of bare or almost bare atomic nuclei, together with vast numbers of free electrons, constitutes a type of gas called a *plasma*. Stars are, in fact, plasma balls. So, in order to understand how a star works, we need to investigate the properties of a hot gas itself, and this means having a look at the part of physics that deals with heat and temperature. In fact, we'll see in this chapter how it will enable us to understand, for example, how a Cepheid variable star pulsates and why enormous convection currents exist within the envelopes of red giant stars, and of course it will also tell us why most stars, including our own Sun, remain stable for very long periods of time. First, though, let's establish a few basic facts about gas.

A volume of gas such as exists within a room will contain a vast population of atoms and molecules and, as just stated, a volume of gas within a star will consist of atomic nuclei and free electrons. So, in any gas, we're talking about atomic and subatomic particles whose physical dimensions are very tiny compared with the volume that the gas occupies. The fact is that in a gas, these particles are free to move around and to cover distances that are large compared with their size.

By contrast, the movement of particles that make up a liquid is more restricted, and for a solid, it is very restricted, simply because the particles are spaced much closer together. The simple fact that the particles in a gas do move around means that they have energy; physicists call this type of energy *kinetic energy*, and in fact, the study of the physical properties of a gas is often referred to as the "kinetic theory of gases."

In any volume of gas, the particles will have a range of speeds, but the average speed is what effectively fixes the temperature of the gas; the higher the average speed the higher the temperature. So the temperature of a volume of gas is really a measure of its energy content. This idea is formalized in what's known as the *first law of thermodynamics*, which essentially just says that heat is a form of energy. This form of energy makes the inside of a star a very interesting place, but in order to appreciate this, we shall do some thought experiments using just the above basic facts, together with a simple cylinder of gas.

From a Cylinder of Gas to the
Structure of a Star

Our "equipment" consists of a cylinder containing the experimental gas, with a movable piston that, though completely frictionless, is nevertheless completely gas tight, as shown in Fig. 1.

piston

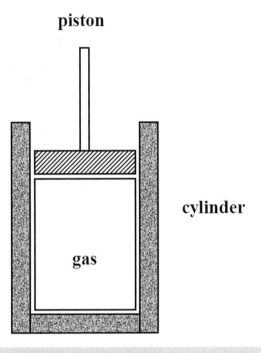

cylinder

gas

Figure 1. The "standard apparatus" used to explain basic processes in the physics of heat and thermodynamics – a cylinder containing gas together with a movable piston that is completely frictionless but is also perfectly "gas tight."

The first thing to note is that the gas within the cylinder must be supporting the weight of the piston plus the weight of the outside air above it. If it did not, then clearly the piston would fall further down the cylinder. What in effect keeps it suspended is the combined effect of a huge number of gas particles, constantly colliding with and rebounding off the piston's underside. One particle, despite moving with what may amount to a pretty high speed, doesn't deliver much of a "punch," but multiply this many billions of times, and the gas in our cylinder can exert enough force to keep the piston in its place; this process is shown schematically in Fig. 2.

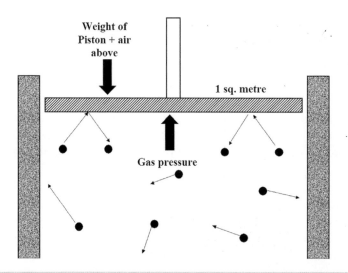

Figure 2. The vast population of atoms in a volume of gas are constantly colliding with their surroundings, which results in the gas itself exerting a force called the gas pressure, or just simply the pressure. This is defined as the force exerted on each unit surface area of the surroundings. Here the pressure of the gas in the cylinder supports the weight of both the piston and the air above it.

If the surface area of the piston happens to be one square meter, then the force exerted by the gas on it is called the *gas pressure*, or just the pressure. One thing that immediately follows from this is that if we raise the temperature of the gas, then the particles would be moving around with higher speeds, which would result in more force, i.e., a higher pressure being exerted on the piston. So increasing the temperature of a volume of gas will increase the gas pressure.

For our next experiment, we shall compress the gas, and we could do this, for example, by adding some extra weight to the piston. The gas pressure in the cylinder can no longer support all of this weight, and so the piston falls, and as it falls, the particles in the gas will pick up some energy from the moving piston. This raises the average speed of the gas particles and so the gas temperature increases; this in turn raises the gas pressure.

Something else is happening, too; by compressing the gas, we are confining the same number of gas particles to a smaller volume. In other words, we have increased the *density* of the gas. This results in more "hits" by gas particles against the underside of the piston in a given interval of time, which again results in the gas exerting a greater pressure. So raising the density of a gas also increases the pressure. Eventually, as a result of the increased pressure, the gas can support the extra weight, whereupon the piston stops falling. Our conclusion, then, is that if you compress a gas, the temperature, the density, and, as a result, the pressure all increase.

The reverse operation is to remove the extra weight on the piston. Our gas now has more than enough pressure to support just the piston plus air, and so the surplus pushes the piston back up the cylinder. In this situation, work is being done by the gas in moving the piston; doing work means using energy, and this is supplied by the motion of the gas particles. The result of using energy is that the gas particles lose energy; they slow down, and so the gas temperature falls. In addition, the density falls as the volume of the gas increases. Our next conclusion, then, is that if a gas expands, it cools down and the pressure falls. These experiments are shown schematically in Fig. 3.

We'll be coming back to our cylinder of gas several times, but what we know now enables us to understand at least in a qualitative way the equations of stellar structure. But because we won't be dealing with actual "equations" as such, we can also refer to them here as the principles of stellar structure.

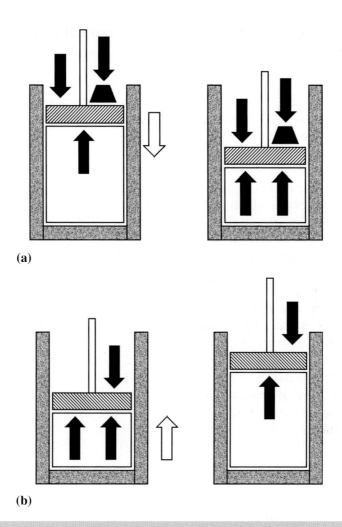

(a)

(b)

Figure 3. In **a**, we add an extra weight to the piston. The gas pressure in the cylinder can now no longer support the weight, and so the piston falls. This increases both the temperature and density of the gas, which results in increased gas pressure, and the piston stops falling. In **b**, we remove the extra weight; the gas now has a surplus of pressure, which forces the piston back up. The gas uses energy to do this; it expands, the temperature, density, and pressure all fall, and the piston stops rising.

The Principles of Stellar Structure

A star consists basically of three parts: the core, where as we shall see later *all* of the star's energy is generated; the envelope, which makes up the bulk of the stellar material; and the surface layers, particularly the photosphere, which is the only part of a star that can be directly observed. One of the very first things that we tend to learn about stars is that for the most part they are stable and that this is the result of a more or less perfect balance between gravity, which is holding the star together, and the thermal energy, which makes the star shine, trying to blow it apart. The correct physics term for when two seemingly opposing forces are in balance like this is that they are in *equilibrium*.

So a stable star is in a state of equilibrium, as shown schematically in Fig. 4. The equations, or in our case, the principles of stellar structure, essentially put the flesh on the bones of this basic idea. In doing so, they describe the basic physical properties of stars and how these properties change, as we go from the surface layers to the very heart of a star. In the hands of the professionals, these equations are used to construct theoretical models of stars in order to predict their "observational properties." These can then, of course, be compared with real observations, which in turn test how good the models are. They are also used to study the way in which stars evolve. Most amateur astronomers are not likely to have an urgent need to do serious work on the equations of stellar structure, but an understanding of what they tell us is clearly a valuable insight into the way that stars work.

There are four main equations (principles) of stellar structure, and these involve the important numbers of the kind that we have already talked about – namely, the temperature, mass, luminosity, and also the pressure within a star. In particular, they describe how these important quantities change as we move from the stellar core to the surface layers. There are also three sorts of supplementary equations that, among other things, give further information on the pressure within a star.

To avoid having to deal with the inevitable complications (these are always left for the really smart guys, anyway), the stars that we are dealing with are assumed to be perfectly spherical and non-rotating; adopting this simplification means that the important quantities depend only on the distance from the center of the star. Here, then, are the most important things that you need to know about stable stars.

1. *Continuity of mass* – This encapsulates the fundamental premise that gravity holds a star together, and, being an attractive force, it means

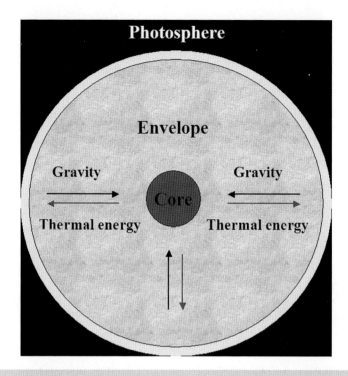

Figure 4. The parts of the "basic star," which is in a state of stable equilibrium between gravity trying to make the star collapse and thermal energy trying to push it apart.

that aside from local minor irregularities, the mass of a star is concentrated toward its center. One obvious consequence of this is that the density of the stellar material decreases, as the distance from the center increases, and what's more this is a steady decrease. This *density gradient* is described more precisely by the mathematical form of this principle.

As an example of the kinds of numbers that we are dealing with, the average density of the Sun is around 1,400 kg/m^3 – almost one and a half times the density of water. The density gradient equation, however, shows that this ranges from around 1.6×10^5 kg/m^3 at the center to more or less zero at the top of the photosphere. Other types of stars will, of course, have a different density distribution, but central densities will be very large for all stars. It often turns out, though, to be a good approximation to assume that the entire mass of a star is concentrated at the center.

2. *Hydrostatic equilibrium* – This is the one that explains in more detail the balance between the gravity and the thermal energy, and we ourselves can understand this better by returning to our cylinder of gas.

This time, we have a modified cylinder of gas, which has two pistons and thus two compartments containing gas, which we have called "gas 1" and "gas 2," as shown in Fig. 5a. Remember, the pistons are totally gas tight, so that these two volumes of gas are isolated from each other. We've also called the gas above the cylinder as "gas 3." The pressure of gas 2 supports the weight of the upper piston plus that of gas 3, whereas the pressure of gas 1 has to support the weight of both gas 2 and gas 3, plus the weight of the pistons. This is more weight, and so clearly the pressure of gas 1 has to be greater than that of gas 2.

Now have a look at Fig. 5b; instead of a cylinder of gas, we have a column of gaseous stellar material, extending all the way from the stellar core to the star's surface. The two pistons are gone, of course, and while the gas within the volumes, which we have again labeled as gas 1, gas 2, and gas 3, might well be free to mix, remember we're dealing with a stable star here, and this means that from one moment to the next things essentially do not change. This, in turn, means that if we take a "snapshot" of the situation at some particular moment, then we effectively have three separate volumes of gas.

As with the cylinder experiment, gas 1 has to support the weight of both gas 2 and gas 3, whereas gas 2 only needs to support the weight of gas 3. So clearly the pressure at the base of gas 2 must be greater than that at the top of gas 2 in order to maintain this state of equilibrium. Another way of stating this is to say that there is a *pressure gradient* in the radial direction, running from the center of the star to its surface. It is this pressure gradient which keeps the star inflated (by exactly the right amount), to prevent the star from suffering gravitational collapse.

As with the previous principle, the corresponding equation describes the pressure gradient more rigorously. Astronomers can use this idea to show that the pressure at the center of a star such as the Sun is at the very least around 5×10^8 times the atmospheric pressure here on Earth.

On a final note here, for most stars, the thermal energy that supports the star comes from the gas pressure, i.e., the kinetic energy of the particles in the stellar material. For very hot stars, a significant contribution to the thermal energy comes from the energy of the radiation itself. Remember from chapter *From Light to Starlight* that photons carry energy. This energy enables the radiation field within the star to exert its own pressure, which not surprisingly is called the *radiation pressure*.

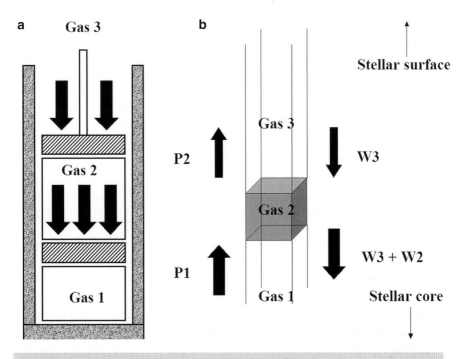

Figure 5. This illustrates the principle of hydrostatic equilibrium in a stable star. On the left, we have a cylinder with two gas compartments separated by two pistons. Gas 1 has to support more weight than gas 2, and so the pressure here must be greater. On the right, we apply this idea to a volume of gas in a stable star. The pressure P2 at the top face of gas 2 supports the weight (W3) of the column of gas above it, whereas the pressure (P1) at the bottom face has to support this plus the weight (W2) of gas 2. For this situation to remain stable, P1 must be greater than P2, and in fact there has to be a pressure gradient running not just through the volume of gas 2 but through the entire star from center to surface.

3. *Thermal equilibrium* – This is necessary to (if you'll pardon the pun) prop up the principle of hydrostatic equilibrium. The most obvious thing about stars is that they shine, and in so doing radiate energy. It is this energy that maintains the aforementioned pressure gradient, and because it is ultimately lost in space, it has to be continually replenished in order to maintain the status quo.

Recall that the Sun is losing energy at a rate of almost 4×10^{26} W; this tells us immediately that the Sun must also constantly generate energy at this rate, and incidentally the figure itself points the way in finding what kind of energy source can do this. As we shall see later, physical conditions such as density and temperature within stars are such that all of the energy is generated within the central core; there is no actual energy production within the greater part of a star's envelope. This means that the luminosity of the star gradually builds up, as we move outwards from the center of the stellar core, but then levels off as we move through the envelope. Once again the corresponding equation of stellar structure gives a more rigorous form to this *luminosity gradient*.

4. *Energy transport* – In order for the energy production within the core of a star to do its job, the energy has to be able to get out through the envelope to maintain the pressure gradient and thus maintain stability.

What actually drives the transfer of energy through a star is the simple fact that a star is hottest at the center and falls (there is a *temperature gradient*) as we move outward toward the star's surface. This principle in equation form describes what determines a star's temperature gradient. The surface temperature of the Sun is the familiar 5,800 K at the photosphere, but rises to around 15 million degrees at the center.

There are three processes by which energy, which is generated in the stellar core, can move through the envelope to the photosphere, and the fact is that whichever of these processes proves to be the most efficient means of doing this will be the dominant process. Thus *conduction*, which works particularly well on many solid materials such as metals (as you'll unpleasantly discover if you leave a metal spoon in a pan of boiling water and attempt to pick it up some minutes later with your bare hands), has very little part to play within the gaseous "world" of stellar interiors. The exception is in white dwarf stars and neutron stars, where densities are exceptionally high, resulting in almost no free movement of atomic particles, except for electrons in their outer layers. It is the movement of free electrons in metallic solids that transfers heat energy in the same way as it does in these extreme stars.

Transport of energy by *radiation* is just what it says; the energy that a star generates is initially in the form of gamma radiation. These high-energy photons gradually work their way through the stellar envelope by being absorbed, re-emitted, and scattered countless times by the particles in the stellar plasma, particularly by electrons, because these very much outnumber the atomic nuclei. These kinds of photon–particle interactions, which result in free electrons (i.e., they are not bound within atoms) gaining or losing energy, are called *free–free transitions*. At every photon–particle event, the original high-energy photon is gradually broken down into increasing numbers of low-energy photons, which finally emerge from the photosphere an estimated 30,000 years after their parent gamma ray was produced. The sunlight that warms your day is very old indeed.

Within a star, where energy transport is by radiation involving free–free transitions, there is a very simple formula for the opacity κ of the stellar material, which is shown in Equation (1). We are already familiar with this quantity from chapter *The Photons Must Get Through – Radiative Transfer* where, you will remember, it is used to calculate how much radiation is absorbed and how much survives to pass through intervening material.

$$\kappa = \kappa_0 \times \frac{\rho}{T^{3.5}} \tag{1}$$

Here κ_0 is just a number whose value depends on the composition (i.e., the presence and quantity of heavier elements) of the stellar material; ρ(rho) is the density of the material, and T is the temperature. The important thing about this formula is that while opacity increases with increasing density, as one would expect, it is in fact very much more critically dependent on the temperature. The fact that the equation involves the temperature raised to the power 3.5 means that a relatively modest rise in temperature will result in a significant drop in the opacity. This equation, which is called Kramers' law, named after the Dutch astronomer Henrick Kramers who derived it in 1923, will play a key role when we look at how some stars pulsate.

Energy transport by radiation works best and is indeed the dominant process in stars when the temperature gradient through the star is not too great and also provided that the stellar material is not so opaque as to seriously inhibit the steady flow of radiation through it. Both of these things can happen in stars, and when they do, radiation simply cannot shift the energy fast enough. The result is that the third energy transport mechanism, *convection*, becomes dominant. This involves large volumes of gas becoming less dense than their surroundings, until they become buoyant and rise upwards, whereupon they cool off, become more dense, and sink back down into the depths. Convection turns out to be a horrendously complex process to describe and model mathematically, but among other things, it almost certainly holds the key for understanding red giant stars. Later, we'll have a very elementary look at convection (aided again by our trusty cylinder of gas).

These, then, are the four main principles or equations of stellar structure; in mathematical form, they basically describe how mass, pressure, luminosity, and temperature change as we move from the center of a star to the surface, and this is a good place to show graphically how these fundamental quantities vary, for example, inside a star such as the Sun, as shown in Fig. 6a–d.

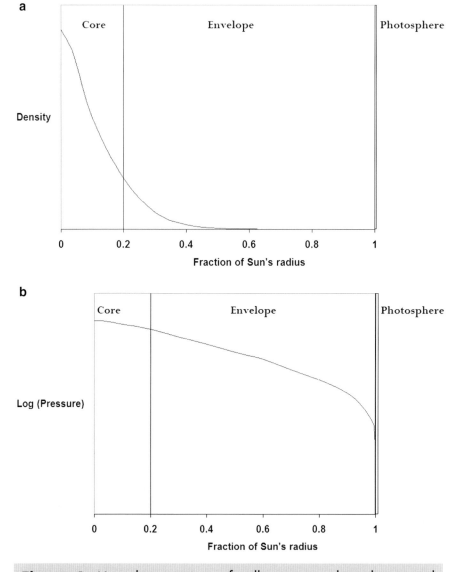

Figure 6. Here the equations of stellar structure have been used to show how the density, pressure, luminosity, and temperature change from the center of the Sun to the photosphere. The important thing to note is that all of these plots show a smooth, steady variation, with no anomalous bumps or wiggles.

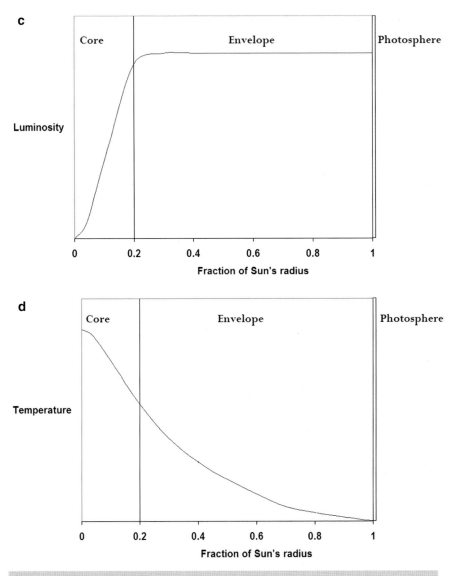

Figure 6. (continued)

There are, in addition, three "supplementary" equations, which give the actual value of the pressure within a star, in terms of the temperature, density, and also the chemical composition of the stellar material. For the most part, stellar material consists of hydrogen and helium with just a trace of heavier elements. Even so, the degree to which hydrogen and helium alone are ionized significantly affects the gas pressure. This particular equation is usually referred to as the "equation of state."

The second of these equations gives the opacity or the extinction coefficient κ of the stellar material – in other words Kramers' law, which we introduced above. This equation is crucial for describing how radiation travels through a star, and the real difficulty for professional astronomers here is that, in order to achieve the "big picture" in constructing theoretical models of stars, it is necessary to know the value of κ for every wavelength. This involves in each case determining the value of κ_0 in Equation (1), which in turn depends on the chemical composition of the stellar material and also on the presence and abundance of different ions. The final equation basically gives the quantity of energy that is generated by each kilogram of stellar material every second, and this is really all about what's going on at the stellar core.

The equations of stellar structure serve among other things to explain why stars remain stable for long periods of time. However, as any astronomer, amateur or professional, knows, the interesting stars are those that are not stable, and there are many processes that can make a star unstable. Some of these involve the envelope, while others involve changes within the stellar core. We'll look at changes in the core in chapter *A Star Story – 10 Billion Years in the Making*, but here we'll look at how disturbing one of the equations of stellar structure can make a star interesting. First, though, it's back to our cylinder of gas.

From a Cylinder of Gas to Stellar Pulsation

When we added extra weight to the piston in our cylinder of gas, the piston fell, compressing the gas, but as we'll now see, there's a little bit more going on in this experiment, and here the details are everything. We add the weight to the piston and the piston falls, compressing the gas; the energy picked up by the gas atoms, together with the increased density, enables the gas to exert a greater pressure on the piston, until it is able to support the total weight of the piston plus extra weights. However, in addition to simply supporting this new load, the pressure exerted by the gas has to actually bring the piston plus weights to a complete stop, and this results in the gas being compressed just an extra bit more.

When the piston *has* stopped, the gas finds itself with an extra bit of pressure – enough to push the piston back up a little bit higher than its rest position. The gas pressure has now dropped a little below that needed to simply support the weight of the piston, so the piston comes back down again until the increasing gas pressure stops it again, and so this goes on effectively ad infinitum. In other words, the piston oscillates about the position where it would be if the gas pressure inside the cylinder were just supporting its weight.

What has happened is that the actual motion of the piston, irrespective of its weight, has given that extra bit of energy to the gas, and the resulting addition to the gas pressure enables the gas to make the piston oscillate. In other words, the gas inside the cylinder is itself now oscillating about a mean value for its density, its pressure, its volume, and even its temperature; it is pulsating. This sequence of events is depicted in Fig. 7a–f.

Finally, if we removed the extra weights on the piston, the same kind of thing would happen. The pressure of the gas would push the piston back up the cylinder, but the motion of the piston would carry it a bit higher than where the now reduced gas pressure could simply support the piston's weight, so the gas is now kind of "overstretched." The piston thus drops and ends up a little below the *equilibrium position* and once more starts to oscillate.

In this more detailed experiment, we have ensured that our cylinder/piston apparatus has been a perfect insulator, which means that no heat energy is allowed to seep through the walls of the cylinder or the piston. Remember, in addition, the motion of the piston within the cylinder

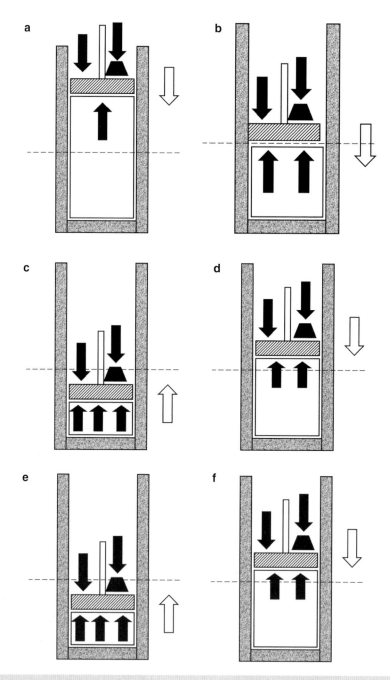

Figure 7. This sequence shows in more detail what happens when we add extra weight to the piston in our cylinder of gas. In each case, the horizontal dashed line represents the position where

is perfectly frictionless, so that no energy can be wasted in making the piston scrape against the cylinder walls.

Finally, we have assumed that the very gas atoms themselves are perfectly elastic. This means that when they collide with the walls of the cylinder and against the face of the piston, they do not lose energy in heating up the material out of which the apparatus is constructed. In this idealized thermodynamic universe, where we can have no transfer of energy either into or out of the volume of gas, the compressions, expansions, and pulsations the gas undergoes are called *adiabatic*.

In the real world, which of course includes the real "world" of a stellar envelope, there will inevitably be some flow of energy into and out of the system, and the result is pulsations that are *non-adiabatic*. In this case, the pulsations of a volume of gas would not continue forever, but instead, as the gas gradually lost that extra bit of pressure due to heat leaking out of the system, the oscillations would subside or, as they say in the trade, would be *damped*.

There is, however, a way to "cheat" the real world and make the pulsations go on forever. We accept now that the walls of our cylinder do allow heat to pass through, and so what we do, just at the moment when the gas in the cylinder is fully compressed, is to briefly apply some heat to the gas, and what's more, we do this each time the piston comes down to its lowest level. In this way, we are repeatedly giving the gas that extra bit of "push," or pressure, to compensate for the losses and to keep the piston going up and down. This process is called "exciting" the pulsation. This is exactly how a diesel engine works. In this case, though, rather than heating the cylinder from the outside, fuel vapor in the cylinder ignites at the moment of maximum compression, which raises the temperature of the gas and in doing so increases the pressure within the cylinder.

Finally, there is also another way in which we can at least partially cheat the system, and this is to make sure that any compressions and expansions

Figure 7. (continued) the piston would be if the gas pressure were supporting it and the extra weight, when everything was stationary. The downward motion, or the *inertia* of the piston, etc., results in the gas being compressed that bit extra, with a resulting slight excess pressure. This excess pressure enables the piston to oscillate or pulsate indefinitely, provided there are no energy losses out of the system.

of the gas happen as quickly as possible. This means that there is less time for energy to leak out of the system, thus keeping things as adiabatic as possible.

Now let's see if we can use what we know to understand how a Cepheid variable star can pulsate.

Cepheid Variable Stars – The Other Story

Cepheid variable stars (named after their prototype δ (delta) Cephei) are most famous due to the work originally done in 1908 by Henrietta Leavitt at the Harvard College Observatory. Her work showed that the longer their period of light variation, the more intrinsically luminous these stars were. This "period–luminosity relation" led, among other very great things, to their use by Edwin Hubble to determine the distances to nearby galaxies (showing that galaxies were indeed "extragalactic") and ultimately for revealing the expansion of the universe itself. The rest, as they say, is cosmology.

Unfortunately, this has, to some extent, resulted in the very nature of what causes the Cepheid (or the "classical Cepheids," as they are often referred to) and related types of variables to actually vary – usually playing "second fiddle" – but not here!

Cepheids are stars that vary in brightness by typically about 1.0 mag. Their light curves are often, but not always, notably asymmetrical, having a sharp increase in brightness followed by a more leisurely decline with, in some cases, a "hump" on the descending part of the light curve. The variability periods for Cepheids range from about 2 to 60 days, peaking at around 6 days; the period of δ Cephei itself is 5.366 days. Schematic light curves for Cepheid variables are shown in Fig. 8a, b.

The very first Cepheid variable star to be discovered was actually η (eta) Aquilae by the English astronomer Edward Piggot in the year 1784. He and his young associate John Goodricke were making a systematic search for new variable stars. (Prior to this time, only a small handful of variables were known.) Indeed, it was later that same year that Goodricke discovered δ Cephei itself and also the famous eclipsing binary star β Lyrae. Goodricke was actually the first person to correctly interpret the light variations of the famous second magnitude star Algol as being due to the mutual eclipses of two stars of unequal brightness, which were in orbit around one another. The fact is that throughout the 19th century and indeed well into the 20th century, it was generally believed that the variations of stars such as δ Cephei and η Aquilae were also the result of them being eclipsing binary systems.

One important observation that appeared to support this view was the fact that absorption lines in the spectra of these variables oscillated regularly about their mean position with a period that exactly matched their periods of light variation. This was interpreted as being due to the Doppler effect (see Appendix 3 in this book for a brief account of how the Doppler effect works), which caused the lines to be slightly shifted

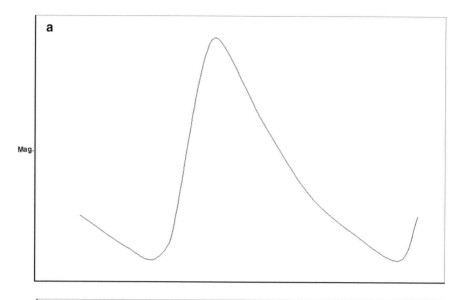

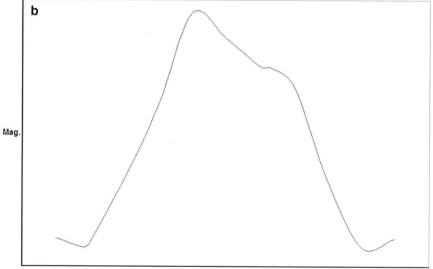

Figure 8. Stylized light curves for Cepheid variable stars, showing their often asymmetric shape and possible "hump" on the declining part of the *light curve*.

alternately towards the blue and the red ends of the spectrum as the visible star swung around in its orbit (and in so doing moved alternately towards and away from the observer). In fact, therein lay one of this theory's problems. There are many binary stars, eclipsing or otherwise, whose component stars are too close to be seen separately in a telescope but which show themselves as binaries by having a "double" spectrum, actually, a single spectrum but with two sets of absorption lines. These are called *spectroscopic binaries*. As the two stars orbit one another, each pair of lines alternately separates and merges together.

There are, to be sure, many such systems where only one component's spectrum is visible, due maybe to the other star being relatively faint, but one would have expected that among the Cepheid-type variables there ought to have been some that showed double spectra. But there were none. In the late 19th century, August Ritter had proposed the idea that a star could pulsate, or alternately expand and contract, and certainly the regular movement of a star's photosphere towards and away from the observer would explain the corresponding shifts in the spectral lines.

Perhaps it was because Ritter was a professor of mechanics, rather than an astronomer, or maybe the very idea that an entire star could expand and contract just seemed too fantastic; the fact is that hardly any notice was taken of Ritter's work. It wasn't until 1914, when the great American astronomer Harlow Shapely effectively "buried" the eclipsing binary theory (though research papers on the idea continued to be published into the 1930s) by pointing out that Cepheids were giant stars and that for some of them to have the observed light variation periods as a result of eclipses, the second star would have to be moving within the body of the main star. The idea of stars being able to pulsate then began to be taken seriously, particularly by the English astronomer Sir Arthur Eddington.

How to Make a Star Pulsate

Ritter's idea was that stellar pulsation involves an outward and inward motion of the entire star. This kind of motion, directed along lines running from the center of the star to the surface, is called *radial pulsation*. Such a pulsation would have to start with some disturbance at the star's core – say a drop in the rate at which heat is generated there. The result would be a drop in the core temperature, accompanied by a drop in the core's gas pressure. This would have exactly the same effect as adding extra weight to the piston in our cylinder of gas, causing the core to shrink and in so doing become hotter. This, in turn, would cause expansion of the core, resulting in cooling yet again, and so on.

This radial oscillation of the core would actually have a relatively small amplitude, because in this part of the star, the density of the gas is extremely high, and so even with a lot of energy involved, there is a lot of material to shift. As the oscillation spreads outwards through the star, very much in the manner of a longitudinal wave consisting of compressions and rarefactions, the density of the gas would fall, and as a result, the amplitude of the oscillations would increase considerably. Eddington, in fact, showed that further out in the envelope of the star, the oscillation becomes increasingly non-adiabatic, and the resulting dissipation of energy simply cannot be compensated for by what's going on at the stellar core. The increasingly damped oscillation basically fizzles out, which simply means that a star cannot pulsate in this way. Time for Eddington's plan "B" then!

Plan "B" was, in Eddington's own words, "fantastic for a thermodynamic engine, but not necessarily for a star." It involved a layer further out in the envelope of a star that could alternately trap and release heat – in other words, the "driving engine" lies in the stellar envelope rather than in the core. This indeed proved to be the way forward, but ironically it was ultimately rejected by Eddington himself, in favor of his original central "driving engine" idea, possibly as a result of the fact that, by the late 1920s and 1930s, it was realized that nuclear reactions are the source of a star's radiation. Powerful as this energy source might be, it was nonetheless shown that it was still inadequate to drive the pulsation of an entire star.

Yet Another Gas Cylinder Experiment

Imagine a version of our piston and cylinder apparatus in which the walls of the cylinder are perfect insulators. The piston and the bottom of the cylinder, however, are made of a transparent material through which radiation can pass freely, and in addition we have a source of electromagnetic radiation (i.e., a very bright light source) beneath the bottom of the cylinder. Under normal conditions and with the gas and piston in perfect hydrostatic equilibrium with their surroundings, the flow of radiation through the gas will ensure that this situation is maintained.

Now we compress the gas inside the cylinder in order to attempt to make the gas pulsate; this would raise the temperature of the gas inside, but recalling Kramers' law, the opacity of the gas would drop dramatically. The result would be that radiation would pass even more readily through the gas and in such a highly non-adiabatic system, any potential pulsation would be severely damped and would quickly die away; hydrostatic equilibrium would thus be restored. However, with the right kind of gas, something else can happen.

If the initial temperature inside the cylinder is such that the gas is just on the brink of becoming ionized, then the energy that compresses the gas actually goes into ionizing it rather than raising the temperature – *so the temperature hardly changes*. The density still increases, of course, but without the rise in temperature. Kramers' law tells us that this time, the opacity of the gas increases, and this results in a buildup of heat, as radiation comes in through the bottom of the cylinder. In addition to this, because the gas has been compressed, i.e., the distance from the bottom of the cylinder to the piston has decreased, the result is an increased temperature gradient between the bottom of the cylinder and the piston, and this results in a higher rate of heat flow into the cylinder. This is equivalent to an injection of extra heat just at the moment of maximum compression, and just as with our earlier experiment, this excites the pulsation of the gas by compensating for any heat losses.

After a while, the buildup of excess heat causes the gas to expand, become neutral, and cool, but the momentum of the expansion takes it to a volume that is greater than that which was simply required to maintain hydrostatic equilibrium. The gas now finds itself "overstretched," and the piston falls, its momentum carrying it below the equilibrium level. The gas becomes ionized again, the heat builds up, and the cycle starts all over again. We have a volume of gas that is pulsating and whose pulsations are being maintained this time by the properties of the gas itself.

Exciting Stellar Pulsation

We don't have cylinders of gas in stars, of course, but we can think of a star as being built from a series of spherical concentric shells, rather like the layers in an onion. The deeper inside a star we go, the hotter and denser the shells of gas become, and this, as we have seen, results in the pressure gradient that maintains hydrostatic equilibrium within the star. The fact is that a stellar envelope is not a placid kind of place, and at any time, some layer of gas can be expected to get compressed by its neighbors. From what we've seen, though, the effect of Kramers' law ensures that such compressions will be sufficiently non-adiabatic to guarantee the speedy restoration of hydrostatic equilibrium – unless, that is, there's a layer of gas within the star that ionizes as it gets compressed.

If this happens, then the layer of gas will act just like the gas in our cylinder in the previous experiment; the opacity will increase, trapping the radiation coming from the next layer down in the star. In stars, this is called the κ -*process* and refers, of course, to the increase in the gas layer's opacity κ. Also, as with the cylinder experiment, the squeezing of the gas layer increases the temperature gradient from the bottom of the layer to the top. This results in even more radiation flowing in from below, which aids and abets the heat buildup; this is called the γ–*process*. So, provided we have a layer of the right kind of gas under the right conditions, it is possible for this layer to undergo pulsations that do not die away but can raise and lower the very outer layers of what we observe to be a pulsating variable star.

The Right Kind of Gas in the Right Kind of Star

The shells or layers of gas in a star will, for the most part, be made up of hydrogen and helium, so what we are in fact looking for in a potentially pulsating star is one that has a shell of hydrogen or helium that is on the point of becoming ionized. Such critical layers of gas are called *partial ionization zones*. Hydrogen has only one electron to lose, so there can only be one hydrogen ionization zone, whereas helium has two, with two corresponding ionization zones. Hydrogen starts to ionize at around 10,000 K; helium becomes singly ionized at around 15,000 K and loses its second electron at around 40,000 K. It is this second ionization of helium with its higher associated temperature that is now believed to be the pulsation-exciting mechanism for Cepheid-type variable stars, but the hydrogen ionization zone (we can include here also the first ionization of helium) does have a role to play, as we shall see shortly.

We can be fairly certain that all stars will contain a layer that is at a temperature of around 40,000 K, except for the very hottest stars, whose photospheric temperatures exceed this value. The issue then is one of where within a star the 40,000 K zone should be to ensure maximum pulsation efficiency, and clearly the lower the temperature of a star's photosphere, the deeper down this layer will be.

If we move along the spectral sequence away from the hottest stars, we'll come to stars whose effective temperatures are actually around 40,000 K, and this initially suggests that the surface layers of these stars might be capable of pulsating. However, in this part of a star, where densities are generally very low, any kind of pulsational disturbance is more likely to permanently eject material from the star rather than induce regular pulsations. The decreasing density, as we head towards a star's outer layers, also shows itself in another more significant way. Basic physics (you can check out the details by looking in any physics book that gives an explanation of simple harmonic motion) tells us that when a volume of gas pulsates (in other words, oscillates about some mean value of volume, pressure, and density), the time it takes for one complete pulsation cycle is directly related to the reciprocal of the square root of the mean density of the gas. So this means that the higher the gas density, the shorter the pulsation period and vice versa. And the closer a layer of pulsating gas is to the stellar surface, the longer the period of pulsation. Get too close to the surface and the pulsation period gets so long that the pulsation effectively peters out.

The bottom line to this is that stars must have low enough effective temperatures so that the helium ionization zone lies at some depth below the surface. With cooler stars, the ionization zone is deeper down, but by going to cooler stars, we effectively run into "red giant land," where it is known that convection is the dominant energy transport mechanism, and this, it is believed, could seriously interfere with stellar pulsations. So the result is an effective temperature zone for stars within which the helium ionization zone resides somewhat below the surface layers but not at too great a depth; we might expect such stars to be capable of sustained pulsation. The place for thinking about stellar temperature zones is, of course, the Hertzsprung–Russell diagram.

Figure 9 shows the location on the HR diagram of the pulsation zone for stars, which is referred to as the *instability strip*. There are three important things to note about the instability strip. First, it runs from the main sequence (involving stars that are somewhat hotter and more luminous than the Sun) almost to the top of the diagram and so involves stars that cover a wide range of luminosities. This, in turn, includes several well-known groups of pulsating variable stars, from the short period δ (delta) Scuti stars, which straddle the main sequence, through to the longer period classical Cepheid variables, which are luminous supergiants.

Secondly, the strip is tilted to the right, which means that the most luminous stars in this zone have lower effective temperatures than the less luminous ones. The third thing is probably obvious, but we'll say it anyway: the instability strip is *not the cause* of stellar pulsation but is the zone where conditions are most favorable for stellar pulsation to be maintained.

From the lower end of the instability strip to the top end, we move from variable stars with relatively short periods (δ Scuti stars have periods of a few hours) to those with longer periods (up to around 60 days for classical Cepheids as mentioned previously). We are also moving from relatively low-luminosity stars to high-luminosity stars, so here we see a visible illustration of the famous period–luminosity relation. The instability strip also tells us something else about this; the higher luminosity stars are giants, which have lower density outer layers than dwarf stars, and as we saw previously, this means that pulsation periods are longer.

So the period–luminosity relation is in fact a period–density relation – straight from the basic physics of an oscillating volume of gas. The right-hand tilt of the instability strip results from the fact that the lower densities in giant star envelopes mean that we need to go to a somewhat lower effective temperature, so that the ionization zone is just a bit deeper and the density is a bit higher for efficiently maintained pulsations.

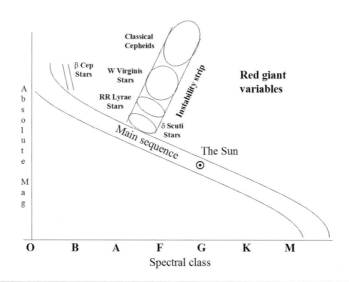

Figure 9. This shows the location of the main groups of pulsating variable stars on the Hertzsprung–Russell diagram. Together they form what is called the instability strip. The left hand or blue side of the strip is determined by the helium ionization zone, not being too close to the surface of the star. The red side is probably less well defined but caused by the increasing dominance of convection as we move to cooler stars. Notice also the less prominent instability strip, which is occupied by the β Cephei group of variables.

On a final note, we've also indicated on Fig. 9a smaller, less significant instability strip right up at the top end of the main sequence where the hot blue luminous stars are. This is the home of a group of pulsating stars called either β (beta) Cephei or β Canis Majoris stars. These are very luminous stars with very short periods and a very small magnitude range. Little is known about their pulsation-driving mechanism, but it has been suggested that it may result from a partial ionization zone, which involves heavier elements. It is also believed that these are stars that are evolving away from the main sequence, and the fact is that the more familiar groups of pulsating stars have also at some stage evolved away from the main sequence – more on this in chapter *A Star Story – 10 Billion Years in the Making*. In the meantime, there's a final matter that we need to clear up regarding Cepheid variables.

Details, Details!

When the helium ionization zone in a Cepheid variable star gets compressed, the result is that hydrostatic equilibrium (i.e., one of the equations of stellar structure) in this part of the star is disturbed. It is as if this layer of gas temporarily cannot support the weight of the overlying layers, and so they too squeeze down. But consisting chiefly of ionized hydrogen, the temperature of this overlying gas increases as it compresses. This should result in a temporary increase in the luminosity of the star and a resulting increase in its magnitude, which should peak at the time of maximum compression.

Spectroscopic observations, which effectively determine the radial motion of the star's photosphere, show however, that Cepheids are at maximum brightness when the star's photosphere is moving towards us at its maximum speed. This happens one quarter of a pulsation cycle, after the time of maximum compression, and is known as the *phase lag problem*. It initially caused many problems, but the culprit is the aforementioned hydrogen ionization zone, which of course lies further up towards the stellar surface. As the overlying gas compresses, this layer, too, gets compressed but becomes ionized. This keeps the temperature more or less constant, and as with the helium ionization zone, the opacity increases and thus traps the upcoming radiation, causing a delay in the rise to maximum magnitude – in other words, a phase lag.

As is well known, red giant stars pulsate, sometimes fairly regularly, but often more erratically. The main driving mechanism for the red giant variables is believed to be the hydrogen ionization zone, but what might otherwise be rhythmic pulsating is probably affected by convection and the associated turbulence in the envelopes of these stars.

From a Cylinder of Gas to a Red Giant Star

For most stars, the main mechanism by which energy gets from the core through the envelope to the photosphere is by radiation, but in the envelopes of very hot stars and red giant stars temperature gradients become too high for energy to be shifted fast enough by this method, and so the main process is convection. Also as mentioned, this is an extremely complex mathematical process and currently involves the use of supercomputers to model the exact details of exactly how and also how much energy is transported through the star. So there's a pretty serious limit to what we can do with our cylinder of gas, but we'll nonetheless have a go.

Notwithstanding the fact that our illustrations of gas cylinders tend to give the impression that they are constructed out of armor plating, let's assume that the cylinder is actually made of very lightweight material, but also (for the purpose of our "experiment") capable of withstanding very high temperatures. The cylinder and piston are also "real" in that they allow heat energy to pass through, but as is also the case in the real world, only at some given rate.

Now let's apply some heat to the base of the cylinder, which will warm up the enclosed gas and effectively create a temperature gradient between this and the surrounding gas outside. Provided this temperature gradient is not too great, the steady supply of surplus heat will simply leak away into the surroundings before there is any significant buildup of heat within the gas inside the cylinder. So, essentially, nothing changes, and a system like this is said to be "stable to convection."

Now we apply a serious amount of heat to the base of the cylinder; this time heat is flowing into the cylinder faster than it can dissipate through the cylinder walls. In other words, we have a much greater temperature gradient between the cylinder and its surroundings. The result is that the gas inside the cylinder expands and in doing so becomes less dense than the outside air. The cylinder consequently becomes buoyant and floats upwards; furthermore, if we happened to carry our heat source within the cylinder itself in the form of some bottles of flammable gas and a gas burner, then our cylinder becomes a hot-air balloon. The large temperature gradient results in the system, consisting in this case of the cylinder of gas, becoming "unstable to convection." If large temperature gradients exist within certain regions of stars, then the material that makes up these regions will also be unstable to convection.

Convection in Stars

Convection is an alternative and actually a more efficient means by which heat can flow outwards through a star, but it will only happen if the other means of energy transport, i.e., radiation, can't do the job, and as we now know, this happens when there is a large temperature gradient and consequently the need for a large amount of heat to be transported quickly. In the surface layers of a star such as the Sun, the various processes that increase the opacity in these layers, such as absorption by heavier elements and negative hydrogen ions, reduce the efficiency of radiation as a means of energy transport. The result is that these layers are unstable to convection, and this reveals itself in the form of the "solar granulation." High-resolution integrated light images of the Sun show the photosphere to have a "speckled" appearance, each "speckle" being the top of a roughly 1,000-km-wide convection cell.

There is also an important difference between heat transfer by radiation and heat transfer by convection. When heat is transferred by radiation, the stellar material doesn't move very much from its own neighborhood, but by contrast, convection involves the large-scale movement of volumes of stellar material through the body of the star. This movement is also accompanied by a great deal of turbulent, i.e., non-smooth, flow of the stellar plasma, and this results in the various chemical elements that make up the star being very thoroughly mixed. A very important consequence of this for red giant stars is that convection currents within the envelope can "dredge up" heavier elements from the core and physically transport them to the surface layers, where they can be expelled as part of a stellar wind and ultimately serve to enrich the interstellar medium.

Clearly convection in most stars has much to do with how conditions in the stellar core compare with those in the envelope. Time then to look at what goes on at the very heart of a star.

The Stellar Core – Source of the River of Starlight

It's difficult to imagine these days the sheer frustration that many of the world's leading late 19th century astronomers must have felt in not being able to explain how the stars shine. For example, it was quickly realized that "King Coal," or indeed any other form of chemical energy, could only keep a star like the Sun going for a period of maybe several tens of thousands of years.

A better proposal was put forward by the German physicist Hermann von Helmholtz; this involved the collapse of material by gravity to form the Sun, for example, and the conversion of gravitational energy into thermal energy. This process would give the Sun a lifespan of around 30 million years – and this period of time is still referred to as the Sun's *dynamical time scale*. It is in fact of the same order as the length of time that it would take the Sun in its present state to cool down in the absence of any replenishment of its internal energy – called the *thermal time scale*.

On the face of it, this was perhaps quite reasonable for the time. However, even then, emerging theories of geology suggested that this was simply not long enough by far for Earth, which was believed to have formed around the same time as the Sun, to have evolved to its current form. The key to this mighty problem in fact lay in the discovery of radioactivity and the nucleus of the atom – alas, a little too late for Herr Helmholtz.

Most of us are familiar these days with the fact that most of the mass of any atom resides at the center in the atom's nucleus. There is an immediate and obvious problem, though, and this is due to the fact that atomic nuclei consist of an assembly of particles called protons and neutrons. Neutrons are electrically neutral; they carry no electric charge, but protons carry a positive electric charge, equal in magnitude to the negative charge of the peripheral electrons. The problem is that in an atom, the nuclear particles, or *nucleons*, are in very close proximity to each other, and this will ensure an enormous force of mutual repulsion between the protons, even though this force may to some extent be screened by the presence of the neutrons.

So there clearly must be an even stronger attractive force that can overcome the electrostatic (or "Coulomb," as it is sometimes known) repulsion. The details of this nuclear force were worked out in the mid-1930s by Hideki Yukawa, the first Japanese Nobel prize winner. This force, which holds the nucleus of an atom together, is nowadays referred to as the "residual nuclear force," to distinguish it from the "strong nuclear force," which holds together the quarks and gluons that make up nuclear particles.

The residual nuclear force, which is also called the "Yukawa potential," is enormously strong over exceedingly short distances of the order of the size of atomic nuclei, but decreases very rapidly – much more rapidly than the inverse square law for the Coulomb repulsion, once we move away from the nucleus. The real key to the energy of the stars, though, lies in the fact that the mass of an atomic nucleus is less than the sum of its parts, which means that if particles such as protons can get close enough that the Yukawa potential overcomes the Coulomb repulsion, then this will mean that there is some mass going spare and, as Einstein said, mass means energy.

As mentioned so many times now, stars consist largely of hydrogen and helium, so let's begin our quest to find the power source of the stars by looking at the masses of these two atoms. The mass of a basic hydrogen atom, whose nucleus consists of a single proton, is 1.674×10^{-27} kg, and so four times this mass is equal to 6.696×10^{-27} kg. The mass of a basic helium atom, whose nucleus consists of two protons plus two neutrons, is 6.645×10^{-27} kg. So the difference in mass between four hydrogen atoms and one helium atom is 5.1×10^{-29} kg.

Now we get our chance to use the most famous equation on the planet, in order to make the stars shine. The equation "$E = mc^2$" inputs the mass "m" in kilograms and multiplies this by the speed of light (in meters per second) squared, to give the equivalent energy in joules. The speed of light squared is 9×10^{16}, which when multiplied by 5.1×10^{-29} gives around 4.6×10^{-12} J. So this is how much energy we could get by taking a group of four hydrogen atoms and fusing them together to make one helium atom. One kilogram of hydrogen will contain $1/1.674 \times 10^{-27}$, or around 6×10^{26} atoms, which in turn equals 1.5×10^{26} groups of four hydrogen atoms. This means that from 1 kg of hydrogen, we could get $4.6 \times 10^{-12} \times 1.5 \times 10^{26}$, which equals 6.9×10^{14} J.

Not bad, but how long would this supply of energy keep the Sun going for? The mass of the Sun's core is about 7×10^{30} kg, and if we assume that when the Sun was formed, around 75% of this was hydrogen, then we start with about 5×10^{29} kg of hydrogen. This then should be able to supply about 3.5×10^{44} J. Observations tell us that the Sun radiates at a rate of about 4×10^{26} W, or 4×10^{26} J/s, and the equations of stellar structure tell us that in order for the Sun to remain stable, it has to be generating energy at the same rate. So if we divide the Sun's total energy supply by the Sun's energy production rate, we get the Sun's expected lifetime in seconds, and this turns out to be about 8.75×10^{17}. Finally, dividing this by the number of seconds in an average year (31,557,600) equals around 2.8×10^{10} years.

We've made a few assumptions and approximations here, but clearly the fusion of hydrogen into helium can run a star such as the Sun for around 10 billion years. Geological evidence suggests that Earth formed around 5 billion years ago. So here, we have it – all the numbers add up, and thermonuclear fusion is the key to starlight.

Nuclear Fusion in Stars – The Details

Thermonuclear fusion of hydrogen into helium can supply the energy needs of a star; however, the conditions that are needed for this process to take place on a scale that is sufficient to maintain the conditions of equilibrium in a star are temperatures of the order of several million degrees, together with high densities. The equations of stellar structure have shown that these conditions only exist within the very central regions, or the cores, of stars. The radius of the Sun's core, for example, is about 0.2 of the Sun's radius. Even so, the fact is that the chance of four hydrogen atoms coming together all at the same time to make one helium atom, even within the high-temperature, high-density regime of a stellar core, is pretty well zero. There's also the business of turning two protons into two neutrons, which all in all means that the fusion reaction has to happen in stages.

Probably the main problem with working out the details of thermonuclear reactions in stars is that laboratory experiments simply cannot simulate the conditions in stellar cores. High temperatures can be achieved, but not the extremely high densities, so that laboratory temperatures need in fact to be even higher than in stars in order to make the reactions work. The result is that the "nuclear chemistry" that takes place in stars has been largely worked out by theory, though the really important product of these reactions, namely, the quantity of energy that is released, has to match that which is ultimately observed. There are in fact two sets of reactions, both of which convert hydrogen into helium, and it may well be that in some stars, both of these reactions operate side by side. However, one of them, the carbon–nitrogen–oxygen, or "CNO," cycle dominates in hotter stars, whereas the proton–proton, or "p–p" chain, takes place in moderate to lower temperature stars.

Even so, as already mentioned, the temperatures required for these thermonuclear reactions to take place are in the order of millions of degrees, which means that energy is only generated in the central hottest region of a star – the stellar core. In turn, it is the temperature of the plasma in the stellar core that essentially determines the probability that a particular reaction will happen.

Now, here's a bit of nuclear chemistry. For hydrogen to be hydrogen, it must have one and only one proton in the nucleus, and in turn helium must have two and only two protons. The number of neutrons that these atoms may contain, however, is a bit more flexible, and there is a standard kind of notation to distinguish between what are in fact different *isotopes* of these elements. For example, "plain hydrogen" with just the one proton is written as "^{1}H," whereas hydrogen in *deuterium* form with one proton and one neutron in the nucleus is written as "^{2}H," and so on. The lightweight form of helium with two protons but only one neutron in the nucleus is then written as "^{3}He."

Figure 10 shows the three main stages in the p–p chain; it's called a "chain" because it essentially starts with two ^{1}H nuclei, and two ^{1}H nuclei pop out at the end to start another link in the chain. Notice that the only stage in which energy is actually generated in the form of a gamma-ray photon is the second stage, the other two basically being element-building stages. The first stage also produces a *positron* (e^+) – the anti-particle of an electron, or an electron with a positive electric charge. This will very quickly encounter a normal electron, whereupon the two will annihilate each other to produce another gamma-ray photon.

In addition, this first stage also produces a particle called a neutrino (ν), which is something of a "will o' the wisp" kind of particle. Neutrinos were conceived by physicists (the name was actually coined by the great Italian physicist Enrico Fermi and means "the little neutral one") to account for an energy deficit in the decay of radioactive elements. As far as the p–p chain is concerned, there are also some other reactions that take place that, in addition to producing some energy (around 15% of the Sun's total energy output, in fact, the main p–p chain accounting for 85%), produce more neutrinos and also the elements lithium and beryllium as intermediate byproducts.

The hot star power source, the CNO cycle is more complicated and involves "outside agents," in the form of nuclei of carbon, nitrogen, and oxygen. This in itself means, of course, that these nuclei have to be present in the nuclear core to start with and we shall see in chapter *A Star Story – 10 Billion Years in the Making* how they get there. In themselves, they serve only to enable the conversion of hydrogen into helium to take place, and they emerge from the process in the same form as they entered it, so they are in effect the nuclear physicist's version of what the chemists call a catalyst. Unless you're seriously interested in nuclear physics, the CNO cycle is very likely not the kind of thing to try and remember in all its details, but here they are anyway.

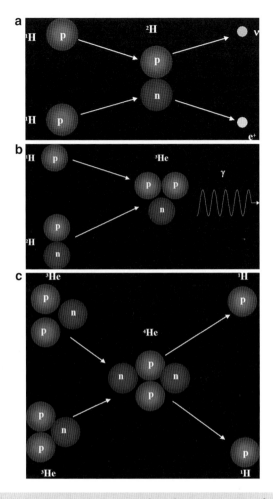

Figure 10. Here we represent the three stages of the proton–proton chain. The second stage generates the energy.

The CNO Cycle

The proton complement for carbon (C), nitrogen (N), and oxygen (O) is 6, 7, and 8, respectively. The other thing to note is that, as the word says, this process is in the form of a cycle, so that there is no "first stage" as such, but any particular stage in the process is followed by the next stage and so on.

$$^{12}C + {^1}H \rightarrow {^{13}}N + \gamma$$
$$^{13}N \rightarrow {^{13}}C + e^+ + \nu$$
$$^{13}C + {^1}H \rightarrow {^{14}}N + \gamma$$
$$^{14}N + {^1}H \rightarrow {^{15}}O + \gamma$$
$$^{15}O + {^{15}}N \rightarrow e^+ + \nu$$
$$^{15}N + {^1}H \rightarrow {^{12}}O + {^4}He$$

The bottom line here is that we now know that stars have a limited energy supply, albeit a very powerful one. In time, the fusion of hydrogen into helium begins to diminish, and this means that all stars eventually change. Stars evolve, and the story of this evolution and its consequences are, as we'll see in chapter *A Star Story – 10 Billion Years in the Making,* some of the most profound things that astronomy has ever taught us. First, though, it's back to our river of starlight and the things that happen to it on its journey across space.

Key Points

- Stars exist in a state of equilibrium between gravity, which would make the star collapse, and thermal energy in the form of gas (and maybe radiation pressure), which would make the star expand.

- The equations of stellar structure enable astronomers to work out in detail how important quantities such as pressure, density, luminosity, and temperature vary from the center to the surface layers of stars.

- A Cepheid variable star pulsates because a disturbance to the condition of hydrostatic equilibrium in the star is maintained by a partial ionization zone, consisting of a shell of singly ionized helium within the star's envelope, which allows heat to build up in the layer. This heat buildup then drives or "excites" the pulsation of the outlying layers of the star.

- The Cepheid "phase lag" is caused by the partial ionization zone of hydrogen closer to the star's surface, which in turn traps upcoming radiation, causing a delay in the star reaching maximum brightness.

- Energy transfer within a star takes place by convection, when the more usual process of radiation becomes inefficient. This can happen, for example, when the temperature gradient within a layer of the star is too large or if the normal flow of radiation is impeded by absorption in cooler layers.

- Convection is an extremely complex process, but it probably holds the key to understanding red giant stars.

- Of all known energy sources, only thermonuclear fusion can supply sufficient energy over a long-enough time scale to enable stars to shine with their observed luminosities and lifetimes.

- Conditions of temperature and pressure for nuclear reactions to take place are such that energy generation only takes place within the central core of a star.

- Most stars fuse hydrogen to helium by the p–p chain, but hotter stars do this by the more involved CNO cycle.

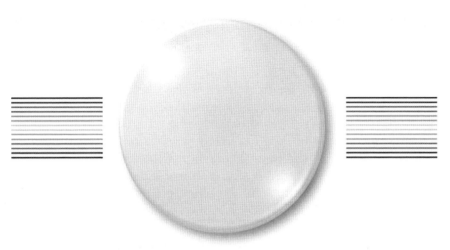

In the Space Between Stars

Another way of defining the photosphere of a star is that it is a layer within the atmosphere of the star, where a photon of a given wavelength has a roughly 50% chance of escaping from the star and heading off into interstellar space. Our "lucky" photon still has a long way to go before it can reach a detector here on Earth, including two major obstacles in its way.

The first of these obstacles is the interstellar medium, which includes anything and everything that lies between the star and the outer layer of Earth's atmosphere. Generally, this is pretty thin stuff, but it extends for as many light years as lie between us and the star, and in fact it's estimated that a star situated close to the center of the galaxy will have its light diminished by 30 magnitudes by the time it reaches us.

The second obstacle is Earth's atmosphere. This extends for only a few miles, but it's pretty thick compared to the interstellar medium. So both of these things can deliver quite a "punch" when it comes to knocking out starlight, but what they actually do to that starlight serves to give astronomers a much clearer picture of the stars themselves. Finally, part of the interstellar medium is in fact "there," because it is not only photons that leave the surface layers of a star.

K. Robinson, *Starlight*, Patrick Moore's Practical Astronomy Series,
DOI 10.1007/978-1-4419-0708-0_9, © Springer Science+Business Media, LLC 2009

Winds from the Stars

The equations of stellar structure showed us that within the envelope of a star, there is a pressure gradient that effectively keeps the star "inflated" and exactly balances the attractive force of gravity, which would otherwise make the star contract – this situation being referred to as hydrostatic equilibrium. Within the body of the star, it serves to essentially keep an individual volume of gas "in its place," but at the very topmost layer of the photosphere, the pressure gradient can enable the most energetic (i.e., fastest moving) particles to leave the photosphere and migrate upward into the star's outer atmosphere – or in the case of the Sun, into the solar corona.

What's more, a pressure gradient must also exist within the corona itself in order to keep it "inflated," and by the same line of reasoning, it follows that particles in the corona's topmost layer can in this case escape the Sun altogether. Thus the pressure gradient in the solar corona drives an outflow of what consists mainly of protons and electrons, which we recognize today as the *solar wind*. The important point here, aside from the fact that the solar wind is driven by the pressure gradient in the corona, is that it results in the Sun continually losing mass. In fact it is interesting to see how long it would take for the Sun to "evaporate" in this way.

Since the 1960s space probes have studied the solar wind extensively, and of course the *Apollo* astronauts set up solar wind detectors on the surface of the Moon. This enables us to give a few facts and figures about the solar wind, which we can use to estimate the Sun's mass loss rate. First, though, here's a useful piece of notation that is used extensively by both physicists and astronomers. When talking about the way in which some quantity *changes with time*, it is standard practice to write the symbol for that quantity with a dot over the top of it. So, for example, the rate at which the radius "R" of a Cepheid variable was changing with time would simply be written as "$\dot{R}$." This notation applies to any quantity whatsoever, but *only* when we're talking about how that quantity changes with time. The rate at which the mass of the Sun $M_\odot$ is changing as a result of the solar wind, i.e., the mass loss rate, can then be written as $\dot{M}_\odot$.

The observational evidence tells us that the density of the solar wind in the vicinity of Earth is equivalent to around 5×10^6 hydrogen atoms per cubic meter. The velocity of the wind particles is very variable and depends on where we are in the sunspot cycle, as well as on whether the particles originate in an active region of the Sun, such as a solar flare or in what might be called a "quiet region." The average velocity is around 400 km/s but can range from around 200 km/s to around 700 km/s.

If we take the solar wind velocity as 400 km/s or 4×10^5 m/s, then imagine a spherical shell 4×10^5 m thick and a radius equal to the Sun–Earth distance, i.e., 1 A.U. or 1.5×10^{11} m. In 1 s, this shell will "fill up" with solar wind particles equivalent to 5×10^6 hydrogen atoms per cubic meter. In the next second the shell will "empty" and fill up again with the same number of particles. So each second, the Sun is losing mass equivalent to the number of hydrogen atoms that fill the shell. The volume of the shell is equal to its surface area multiplied by its thickness (this is actually an approximation, but because the thickness of the shell is very small compared to its radius, it is a very good approximation) or, $4\pi \times (1.5 \times 10^{11})^2 \times 4 \times 10^5$ m^3. With pocket calculator in hand, you'll discover that this is equal to; 1.13×10^{29} m^3. The mass of a hydrogen atom is 1.674×10^{-27} kg and so a density of 5×10^6 hydrogen atoms means a mass of 8.37×10^{-21} kg of hydrogen per cubic meter, and so the total mass in the shell, i.e., the solar mass loss rate per second, is equal to $8.37 \times 10^{-21} \times 1.13 \times 10^{29}$, which equals around 9.5×10^8 kg. Taking an average year as 365.25 days or about 3.2×10^7 s, the annual solar mass loss rate is equal to about 3×10^{16} kg. This seems like an enormously large amount of mass for the Sun to lose each year, but bear in mind that the mass of the Sun is about 2×10^{30} kg; it would take about 10^{14} years (the total mass of the Sun divided by the annual mass loss rate) for the Sun to "evaporate" as a result of the solar wind. This is effectively an eternity, and things will happen to cut the Sun's life short way before this time arrives.

Dividing the annual mass loss rate by the Sun's total mass, we can finally say that $\dot{M}_\odot$ is about 10^{-14} $M_\odot$ per year, or 10^{-14} solar masses per year. On a general note it is standard practice when talking about the mass loss rates for stars that lose mass by whatever means to speak in terms of solar masses per year.

Mass loss rates as low as that in the solar wind would probably not be detectable in other stars, but there are stars that are known to have winds with much higher mass loss rates. However, in these cases, there is a much more efficient wind driving mechanism than simple gas pressure, and this is the pressure exerted by the very radiation itself from these stars. This is called *radiation pressure*. We know that photons carry energy, and it goes without saying that photons "move"; anything that carries energy and that moves can exert a force. In the case of radiation, this force is not surprisingly directly related to our old friend the radiative flux, which we met in chapter *Space – The Great Radiation Field*. In the radiation field, which is present in the outer layers of a star, large numbers of photons and specifically large numbers of high energy or short wavelength photons will result in large flux values and a correspondingly higher radiation pressure. So a good place to look is high luminosity hot stars.

Stars of spectral class O and the even hotter Wolf–Rayet stars are known to have out-flowing winds, with velocities of the order of a few thousand kilometers per second. The main evidence for this is that the spectra of these stars include emission lines of hydrogen, so that there is clearly a very active chromosphere, or layer of hot thin gas, in their lower atmospheres. However, in many cases, there is an absorption line adjacent to and on the blue side of the emission lines, caused by absorption in an outer layer of hydrogen, which is cooler and also moving toward us. A layer of hydrogen, which sits "unmoving" between us and a source of radiation that is producing a continuous spectrum, i.e., a star, will absorb light at the same wavelengths, which are measured in a laboratory.

However, if this layer of hydrogen atoms is moving toward us, i.e., moving away from the light source, then the atoms in the layer will "see" the incoming photons as being red shifted by the Doppler effect. For example, a photon coming from the star that we observe to have a wavelength of, say, 6,555 Å and that we would expect to pass unhindered through the hydrogen layer, could be "seen" by the hydrogen atoms as having a wavelength that is Doppler shifted to 6,563 Å, and so the photon gets absorbed and forms part of what we see as a blue-shifted Hα absorption line.

This combination of an emission line and a blue-shifted absorption line is called a *P Cygni profile*. Mass loss rates from these very hot stars are estimated to range from 10^{-7} to 10^{-4} $M_\odot$ per year, and this will result in these stars losing a significant amount of their mass over the course of their lifetimes – though, as we shall see in the next chapter, the lifetimes of these stars are relatively short anyway.

Stardust

Cool stars also have winds, though in this case the winds are slow moving with velocities of around 10–20 km/s. Red giant stars actually have their own versions of P Cygni profiles in the form of additional blue-shifted components added to the normal stellar absorption lines.

Despite the low wind velocities, mass loss rates in cool giant stars can amount from around 10^{-9} to as high as 10^{-5} $M_\odot$ per year. In the case of these stars, though, the wind material consists not just of hydrogen, etc., but of dust grains, which are believed to "condense" (by a process as yet still not fully understood) in the outer layers of these stars. The dust grains are "pushed" outward by radiation pressure, and it is interesting to note that the radiation pressure acting on the grains obeys a form of "inverse-square law," just like that for gravity. In this sense, the radiation pressure is acting like a form of "anti-gravity," which if it just exactly balances the gravitational pull on a dust grain will keep the dust grain suspended indefinitely in the atmosphere of the star. Furthermore, the state of exact balance occurs (more advanced books will show you how to calculate this) when the ratio

$$\frac{L_*}{M_*} = 6.25 \times 10^{-5} \tag{1}$$

Here L_* is the star's luminosity in watts, and M_* is its mass in kilograms. If this ratio exceeds 6.25×10^{-5} (i.e., increase the luminosity and/or lower the mass), the radiation pressure overcomes gravity and the grain escapes. For the Sun the ratio is around 1.9×10^{-4}, and so any dust grains formed in the outer layers of the solar atmosphere should be able to leave and become part of the solar wind.

The long-term effect of all of this is, of course, that the material expelled from stars feeds out into and becomes part of the interstellar medium.

When the Galaxy Shrank

One of the great astronomical tomes of the 20th century was the work carried out by Harlow Shapely in 1918 on the distribution of globular star clusters. This led to the realization that the Solar System does not lie at the center of the Milky Way, but what at the time appeared to be a distance of around 50,000 light years from the center.

Later, however, work by the Dutch astronomer Jan Hendrik Oort (of "Oort cloud" fame) reduced this distance to around 30,000 light years. The reason for this was that the Swiss American astronomer Robert Julius Trumpler had made a comprehensive study of open or galactic star clusters. Trumpler made the (possibly unreasonable) assumption that these clusters were all of approximately the same actual size and brightness. What made his assumption reasonable was the fact that the more distant a cluster was, the fainter it was, fainter than it should have been. If, in fact, open clusters came in a wide range of sizes and overall brightness, some of the more distant clusters should have appeared brighter than expected.

This systematic extra faintness for distant open clusters led Trumpler to the conclusion that interstellar space contains material that dims starlight, making relatively close objects appear fainter and hence further away than they really are. Further investigation showed that it was light at the short wavelength end of the e-m spectrum that suffered most from this *interstellar extinction*. Oort applied these results to the globular clusters, which have a more or less spherical distribution around the galaxy. This not only "pulled in" the distances to the globular clusters, it also meant reducing our distance to the galactic center in order to preserve the correct globular cluster distribution. The long-term result of Trumpler's work was that astronomers must now understand what happens to starlight as it passes through the interstellar medium, in order that they can arrive at more accurate estimates of those all important fundamental stellar parameters, like luminosity, mass, radius, temperature, and, of course, distance.

Space Dust

In chapter *Space – The Great Radiation Field* we talked about the radiation field, which is made up of all those rivers of starlight that crisscross interstellar space; indeed, this great radiation field is often referred to as "the interstellar radiation field." The rest of the stuff, which occupies the space between the stars, consists of atoms, molecules, and dust grains, which taken together make up the interstellar medium, or "ISM," that, all in all, accounts for about 5% of the "ordinary" matter in the galaxy.

Around 99% of the ISM consists of gas – predominantly hydrogen in the form of "loose" atoms (atomic hydrogen), as well as molecules consisting of two atoms chemically bound together (molecular hydrogen). The remaining 1% is called "dust," though it very likely consists mostly of tiny grains of frozen gases, including grains of water ice, although there are also very likely grains of things such as graphite and silicate materials.

One could be forgiven, then, for thinking that interstellar hydrogen would be the main cause of interstellar extinction. Not so, however, unless we're dealing with a special region of the ISM, such as the Orion Nebula. The Orion Nebula and other emission nebulae like it are special, because they have in their vicinity hot stars that emit high-energy ultraviolet photons. These photons ionize the surrounding hydrogen; recombination follows ionization in an ongoing process, which causes these nebulae to glow in the visible spectrum and also causes the ultraviolet flux from these hot stars to be diminished. The absorption of photons in emission nebulae actually heats up the hydrogen gas to a temperature of around 10,000 K, but away from these special regions, space, as they say, is very cold. This means that hydrogen atoms will have their electrons in the lowest energy level, and the general lack of sufficiently energetic photons in the ambient radiation field means that it is very likely how they'll remain.

Absorption as a whole, in fact, plays a rather restricted role in what the ISM does to starlight. There are within the ultraviolet, visible, and infrared regions of the spectra of some stars weak absorption features called "diffuse interstellar bands," which these days are generally ascribed to complex organic molecules, and there is also a well-known feature in the ultraviolet at around 2,200 Å. This is actually called the "2,200 angstrom extinction bump"; it is believed to be due to grains of carbon, mostly in the form of graphite, but possibly also in part due to Buckminster fullerene, which comes in the form of soccer ball-shaped molecules consisting of 60 carbon atoms. The other mechanism for diminishing starlight is scattering, but once again hydrogen gas is not the cause.

We saw in chapter *The Photons Must Get Through – Radiative Transfer* that scattering, such as that which takes place in Earth's atmosphere, is inversely proportional to the fourth power of the wavelength of the light that is scattered. The daytime sky is blue, and clearly the sunlight has to some extent been reddened on passing through the atmosphere. However, the observational evidence shows that where interstellar reddening is concerned, the extinction is inversely proportional to only the first power of the wavelength – in other words, the wavelength dependence is not as significant as it would be for the scattering of gas atoms.

If light is scattered by particles, such as for example sand grains, whose size is obviously much greater than the wavelength of light, then the amount of extinction is pretty well independent of wavelength. The theory of scattering says, however, that for the extinction to be inversely proportional to the first power of the wavelength, the size of the entities that are doing the scattering must be roughly equal to, in this case, the wavelength of visible light, which puts them in the range from about 10^{-5} to 10^{-4} cm. This is about the same size as tobacco smoke and is much larger than the typical size of atoms and molecules. So the "smoking gun" of interstellar extinction is in fact scattering by the "dust" or the grains in the interstellar medium.

The Dimming of Starlight

The most basic and obvious effect that the interstellar medium will have on a star is to diminish its apparent magnitude compared to what it would be if interstellar space were empty. With what we learned in chapter *A Multitude of Magnitudes for the Colors of Starlight*, we can be much more specific about this. Let's call the U, B, and V magnitudes that a star would have in the absence of interstellar extinction U_0, B_0, and V_0, respectively. The first thing to say then is that as a result of interstellar absorption, the observed values of U, B, and V will all be fainter, that is, they will have larger numerical values; and so $U - U_0$, $B - B_0$, and $V - V_0$ will all be positive numbers. What's more, because interstellar absorption has a greater effect on shorter wavelengths, $U - U_0$ will be larger than $B - B_0$, which in turn will be larger than $V - V_0$. Hence

$$(U - U_0) - (B - B_0) = \text{a positive number} \tag{2}$$

and

$$(B - B_0) - (V - V_0) = \text{a positive number} \tag{3}$$

We can rearrange both of these:

$$(U - B) - (U_0 - B_0) = \text{a positive number} \tag{4}$$

and

$$(B - V) - (B_0 - V_0) = \text{a positive number} \tag{5}$$

Equations (4) and (5) give us the difference between the color index that a star is observed to have and that which it would have if there were no interstellar absorption. This difference in the color index is called the *color excess* for the star; it is always a positive number, which gets bigger with increased interstellar absorption, and it is written in the literature as $E(B - V)$, etc. So

$$E(U - B) = (U - B) - (U_0 - B_0) \tag{6}$$

$$E(B - V) = (B - V) - (B_0 - V_0) \tag{7}$$

The color excess is telling us that interstellar absorption causes the color indices to get bigger; even an initially negative color index for a very hot star can turn into a positive value as a result of interstellar absorption. In effect, stars appear redder than they should, and so the whole business is often referred to as *interstellar reddening*. A better name for it might be "interstellar de-bluing", because it is always the blue end of a star's spectrum that is weakened the most, but perhaps this doesn't roll off the tongue so well.

A good question at this point would be, how do you know whether or not the light from a star has in fact been reddened at all? Put another way, how do you, as an observer, know that your observed values for U, B, and V are not in fact simply the U_0, B_0, and V_0 values for this particular star? The answer lies in the appearance of the star's spectrum, and a good example here would be a star whose spectrum was fairly weak at blue wavelengths (suggesting at first that it is a cool star of, say, spectral class M), but whose red continuum showed a notable absence of the kinds of spectral features one sees in the spectra of cool red stars. These would be many absorption lines due to various heavier chemical elements, and absorption bands due to molecules. This would make you strongly suspect that you are in fact observing a hotter, bluer star that has suffered significant interstellar absorption.

It is clearly important to know what the spectrum of a star would look like in the absence of interstellar absorption, and this was achieved by making careful and detailed observations of nearby stars (as determined, for example, by trigonometric parallax). These observations can be used to calibrate important spectral features, such as the relative intensities of prominent absorption lines for the various spectral classes. This, in turn, enables the true spectral class of more distant stars to be identified, despite the "spectrum altering" effects of the interstellar medium. In addition, these observations enable the true values of the color indices $(U_0 - B_0)$ and $(B_0 - V_0)$ to be determined for the various spectral classes. These true color indices are often referred to as *reddening free* or sometimes the *intrinsic* color indices. This in fact is how the color–color diagram that was introduced in chapter *First Look Inside a Star – The Atmosphere* was produced.

One immediate effect of interstellar reddening, particularly on the hotter, bluer stars (cooler red stars after all don't have quite so much blue light to "give away"), is to shift their positions on the color–color diagram. Figure 1 shows what happens; the effect of increased numerical values for both U – B and B – V is to shift a star's position down and to the right by an amount E(U – B) and E(B – V), respectively, compared to its position in the absence of interstellar reddening. The line that connects these two positions is referred to as the *reddening line*.

A very important result here, which has come from extensive studies of the effects of interstellar absorption, is that the value of the ratio of

$E(U - B)$ divided by $E(B - V)$ has almost the same value of approximately 0.72 for all main sequence stars, which is another way of saying that this *reddening ratio* is more or less independent of both spectral class and distance. This ratio is simply the gradient, or steepness, of the reddening line, as plotted on the color–color diagram; it means, for example, that two stars of the same spectral class but at different distances will lie at different points on the same reddening line. It also means that if you were to make photometric observations of a star using the standard Johnson and Morgan filters, in order to determine the $U - B$ and $B - V$ indices, you could then plot the position of this star on the standard color–color diagram.

Now you draw a line having the correct gradient, so that $E(U - B)$ divided by $E(B - V)$ for the star equals 0.72 through the star's position and where it intersects the color–color plot. This gives you an estimate of the star's un-reddened or intrinsic $U - B$ and $B - V$ colors. Some ambiguity can occur in the vicinity of the "U-shaped bend" of the color–color diagram, as a result of the reddening line crossing the U-B vs. B-V plot at more than one place, and this is resolved by having information on the star's spectral type. Where there is no ambiguity, the intersection of the reddening line with the color–color plot can of course serve to give you an actual estimate of the star's spectral type. So use of the reddening line enables you to make an estimate of a star's true colors, and the estimated value of $B_0 - V_0$ gives you an estimate of the star's temperature, even though the light from the star has been modified by the interstellar medium.

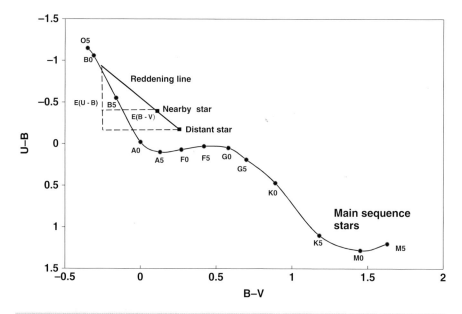

Figure 1. The effect of interstellar reddening is to shift the position of a star downward and to the right of where it would otherwise be on the color–color diagram. The reddening line, which joins these two positions, has roughly the same slope for all stars, which means that two spectroscopically identical stars at different distances will simply lie at different points on the same reddening line.

Through the Atmosphere

The final stage of a starlight photon's journey brings it through Earth's atmosphere to our detector. One of the first things that every amateur astronomer learns is that you should always observe relatively faint things, such as stars and deep sky objects, when they are as high above your horizon as possible, and the reason for this, of course, is that the distance traveled by the photons through the atmosphere is minimized.

Earth's atmosphere dims starlight, just like the interstellar medium, but by how much? After all, when we talk about the magnitude of a star, what we really mean is the magnitude as observed at the top of the atmosphere. However, failing the possibility of becoming an astronaut, we are well and truly stuck with making our observations from the bottom of the atmosphere. So how do we measure a star's "top of the atmosphere" magnitude? The procedure is actually quite cunning, and it originated with the 18th century French mathematician Pierre Bouguer, who is regarded as one of the founders of the general science of photometry; or the basic measurement of the intensity of light, of which astronomical photometry is of course a part.

Let's begin by using a detector such as a CCD camera, or a photoelectric photometer, to observe a star that is exactly at the zenith as seen from our observing location. Our detector will produce an output reading directly related (in a way whose exact details we don't actually need to know) to the flux value that we receive from the star, so we'll allow ourselves to be a bit cheeky and actually call this output reading "the flux." What we do know from chapter *Space – The Great Radiation Field* is that flux ratios, or in this case the ratios of detector output readings, convert directly into magnitude differences. We can also safely assume that the flux value that we measure will be less than that, which would be measured at the top of the atmosphere. This, in turn, means that any magnitude "m" that we derived for the star would be fainter than the corresponding magnitude "m_0" that the star would have as seen from the top of the atmosphere, and we want to know by how much.

As Earth rotates, the star will move away from the zenith; we take another flux reading and also measure the star's zenith distance "z" (the angular distance on the sky between the star and the zenith itself). We continue this process of taking flux readings and zenith distance measurements, while keeping a record of our results. In order to avoid complications (and there are always complications with these kinds of things and here, we are avoiding the complications by adopting the plane parallel atmosphere approximation, just as we did with the Sun's photosphere), we continue our measurements until the star has a zenith distance of about 60° (so there is in fact still plenty of sky to play with). We would of course observe that as the zenith distance increased, the observed flux values for the star decreased, as its light shone an increasing distance through the atmosphere.

At the zenith itself the starlight is normal to the plane of the atmosphere, and so its path length through the atmosphere is simply equal to the height of the atmosphere, which we can call "h." By restricting the star's zenith distance to less than 60°, we can ignore the curvature of both Earth's surface and the surrounding atmosphere. We can also ignore the fact that because the density of the atmosphere increases as one gets nearer to Earth's surface, the paths of light rays, which come from other directions than the zenith, are slightly curved because of refraction.

The opacity of the atmosphere; that is, its ability to absorb and scatter incoming starlight will, of course, vary with height, because the atmosphere gets less dense with increasing height. This varying value of the opacity will combine with the height of the atmosphere to produce an optical depth through the atmosphere in the direction of the zenith.

As we shall see, we won't actually need to know what this zenith optical depth or "τ_0" actually is, so we don't need to worry about the exact way in which the opacity varies with height, and in fact we can simply assume an average opacity "σ" for the atmosphere that, when multiplied by the path length, will give us the optical depth. We do need to assume, though, that the opacity of the atmosphere at any given height directly above our head is the same as that at the same height above the ground in the vicinity of our observing location, just so that we can assume that this average opacity stays the same. As before, the best way to ensure this is to keep the zenith distance within the 60-degree limit, so that points on the starlight's path don't get too far away from our neighborhood. The zenith optical depth in the atmosphere then is simply given by

$$\tau_0 = \sigma \times h \tag{8}$$

Now consider the star as having "moved" to a zenith distance of "z"; the new path length for the starlight through the atmosphere, which we can call "l," is now equal to h divided by the cosine of the zenith distance, as shown in Fig. 2. So

$$\ell = h/\cos z \tag{9}$$

This is, of course, exactly the same formula that we used when talking about the optical depth of the Sun's photosphere. The reciprocal of the cosine (i.e., 1 divided by the cosine) of any angle is called the *secant* of the angle, abbreviated as "sec." Thus

$$\ell = h \times \sec z \qquad (10)$$

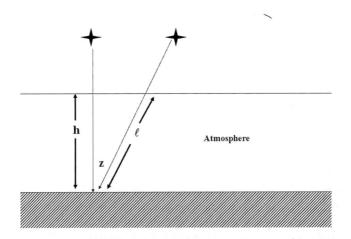

Figure 2. For a star having a zenith distance of "z," the starlight path length through Earth's atmosphere, *l* is equal to *h*/cos(z), which equals *h* × sec(z).

So the optical depth "τ" through the atmosphere for a zenith distance of z is equal to

$$\tau = \sigma \times h \times \sec z \qquad (11)$$

for example

$$\tau = \tau_0 \times \sec z \qquad (12)$$

In chapter *The Photons Must Get Through – Radiative Transfer*, we saw that a beam of starlight with an initial intensity value of I_{in}, which passes through an absorbing medium of optical depth τ, comes out the other side with a diminished intensity that we called I_{out} and whose value is given by; $I_{in} \times e^{-\tau}$ where "e" is the number 2.718. Here the absorbing medium is the atmosphere, and because it covers a vanishingly small distance compared to the distance to any star, we are in fact okay to use flux values rather than intensities in this situation. The optical depth is given by Equation (11), so the flux value "F" that we receive at ground level from the star when its zenith distance is "z" is simply

$$F = F_0 \times e^{-\tau} \qquad (13)$$

F_0 is the flux value that would be received at the top of the atmosphere. The ratio of these flux values is just

$$\frac{F}{F_0} = e^{-\tau} \qquad (14)$$

Now here's another of those simple but very important rules involving logarithms. As an example, the number 5/2 is of course equal to 2.5, and the logarithm of 2.5 is very nearly equal to 0.4. Now use your calculator to give you the logarithm of 5 (i.e., 0.7) and subtract from it the logarithm of 2 (i.e., 0.3); this answer, of course, is the same, i.e., 0.4. This general rule tells us that log (F/F_0) is equal to log (F) – log (F_0).

Remember also from chapter *Space – The Great Radiation Field* we showed that another general rule involving logarithms is that if we take the log of a number such as "$e^{-\tau}$," then this would give us exactly the same number, as simply taking the log of "e" and multiplying it by "$-\tau$." So, if we now take the log of both sides of Equation (14) we'll get

$$\text{Log}(F) - \log(F_0) = -\tau \times \log(e) \qquad (15)$$

The log of "e" is equal to 0.4343, and so with a slight rearrangement

$$\text{Log}(F) - \log(F_0) - \tau \times 0.4343 \qquad (16)$$

We've already seen that "τ" is equal to $\tau_0 \times \sec z$ where τ_0 is the atmospheric optical depth in the direction of the zenith (we don't know the value of this yet) and "z" is the star's zenith distance. So then

$$\text{Log}(F) = \log(F_0) - \tau_0 \times 0.4343 \times \sec z \qquad (17)$$

Our series of observations involves determining the values of F and measuring the corresponding values of z; so for each pair of values we can note down the log of F and the secant of z. We now plot a graph on which log (F) runs up the "y" axis, while sec z runs along the "x" axis. If the universe is being kind to us, our plotted points should follow an approximately straight line, running from the upper left-hand side of the graph to the lower right.

From the general mathematical rules about graphs that are in the form of a straight line, the gradient or the slope of our plot is equal to $-\tau_0 \times$ 0.4343, and the best way to determine this is to mark off two points on the line that are reasonably far apart and form a right-angled triangle, as shown in Fig. 3. The gradient is then just equal to the height of the vertical line of the triangle, divided by the length of the horizontal line. The negative sign for the line's gradient tells us that the plot falls to lower values as we move along the "x" axis. We now divide this gradient by 0.4343, and lo and behold, we've determined the zenith optical depth in the atmosphere and we should be able to tell straightaway how optically thick or optically thin the atmosphere is tonight.

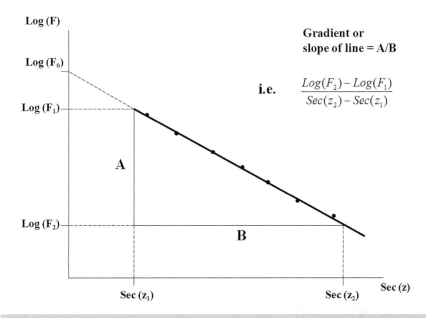

Figure 3. Here we plot the logarithm of the observed "flux" values "F" for a star against the secant of the corresponding zenith distance "z." The plotted points should fall approximately along a straight line, whose slope or gradient is equal to the difference between two of the "Log(F)" values divided by the difference between the corresponding "sec(z)" values. This number divided by 0.4343 gives the optical depth through Earth's atmosphere in the direction of the zenith. The point where the line touches the vertical axis gives the log of the zenith "flux" value for the star.

One of the beauties of this process is that we don't actually need that observation of the star at the zenith; the other observations will still give us the same straight line with the same gradient and the same resulting zenith optical depth. If that isn't enough, if we continue or extrapolate our straight line plot until it crosses the "y" axis, the point of intersection gives us the value of F_0; this is the flux value that would be measured at the top of the atmosphere. Having now derived the value of F_0, we can use it to work out the relative drop in the star's magnitude as it sinks toward the horizon and suffers increasing absorption in the atmosphere. We do this very simply as follows. Equation (5) in chapter *Space – The Great Radiation Field* showed us that the magnitude difference $m_0 - m$ corresponding to a flux ratio of F/F_0 is equal to

$$m_0 - m = 2.5 \times \log{(F/F_0)} \qquad (18)$$

We have F_0 and we have also measured a series of values for "F" at varying zenith distances, so $m_0 - m$ is now the difference between the star's magnitude at the top of the atmosphere m_0, and that at ground level "m." Finally then, how do we get the actual value of m_0?

We know that Vega is the astronomer's "zero star," in that its magnitude in all the various photometric wavelength regions is exactly 0.0. By carrying out our flux reading vs. zenith distance observations on Vega (or indeed on any standard star whose top of the atmosphere magnitude m_0 has already been determined by the unsung efforts of previous astronomers), we can determine the drop in magnitude $m_0 - m$ due to atmospheric extinction for the standard star. It then follows that to estimate the magnitude of a "target star," we can apply the same correction for atmospheric extinction and thus determine its top of the atmosphere magnitude – all in all, a very neat technique from Monsieur Bouguer. The basic principles of the method have been subsequently developed over the years to enable astronomers both professional and amateur to do what's known as "all sky photometry."

There now remains just one story to tell in our introduction to the physics of starlight. This is the story of the stars themselves – how they form, evolve, and eventually "pass away," leaving us a truly wonderful legacy. It is a story that has many twists and turns, but as always we shall keep things as simple as possible, so that we can hopefully keep a clear view of the physics that is involved.

Key Points

- Stars (probably most of them) produce winds that cause them to lose mass.

- The pressure gradient in the outer layers of a star will produce a wind of only a relatively low mass loss rate.

- Winds that are driven by radiation pressure produce much higher mass loss rates.

- Gaseous winds from very hot stars reveal themselves by the presence of P Cygni profiles in the stars' spectra.

- An important component of slow-moving winds from cool giant stars is dust grains.

- The "dust grains" in the interstellar medium are the prime cause of dimming the light from stars.

- Dimming is greatest at shorter wavelengths, resulting in distant stars appearing redder than they would otherwise be, and hence this dimming is called interstellar reddening.

- The amount of interstellar reddening is measured by the difference in a star's color indices, compared to what they would be for a star of the same spectral type and in the absence of reddening. Such a difference is called a color excess.

- One effect of interstellar reddening is to push a star's position downward and to the right in the color–color diagram. The line connecting the star's reddened and its unreddened position is called the reddening line.

- The gradient or slope of the reddening line for main sequence stars has been found by observation to be roughly constant with a value of around 0.72.

- The "top of the atmosphere" magnitude for a star can be found by using Bouguer's method, in which the decrease in a star's brightness is measured alongside its increasing zenith distance.

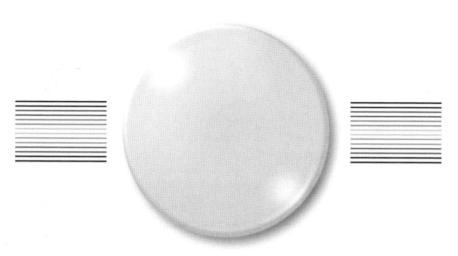

A Star Story – 10 Billion Years
in the Making

These days the story of stellar evolution could almost read like a Hollywood epic – and why not?! It is a story that probably began back at the time of the Renaissance in Europe, when the stars were first suspected as being distant suns. Over the centuries the "script" has been written by countless astronomers, both amateur and professional, some well-known and many unsung heroes. It now stands as one of the greatest of all astronomical tales, equaled probably only by the story of the origin and evolution of the universe itself, and just like that great story, the story of the river of starlight still has a long way to go before we can write "the end."

The story of stellar evolution is one that probably best begins not at the beginning but in the middle, where compared to what comes before and after, life for a star is relatively tranquil. This is the star's time on the main sequence of the Hertzsprung–Russell diagram.

K. Robinson, *Starlight*, Patrick Moore's Practical Astronomy Series,
DOI 10.1007/978-1-4419-0708-0_10, © Springer Science+Business Media, LLC 2009

The Middle of the Story – Life on the Main Sequence

The basic definition of a "star" is an astronomical object that generates energy and emits radiation as a result of thermonuclear fusion taking place in its core. Stars, including their cores, consist mostly of hydrogen, and, as we saw in chapter *Deep Inside a Star*, hydrogen provides a ready source of nuclear energy. Because it is so abundant in stars, it should last for a while, so that stars that are using it to generate energy should remain stable for a long period of time. All stars on the main sequence are generating energy by fusing hydrogen into helium, and they are indeed essentially stable. However, they will only stay on the main sequence for as long they can maintain their energy production rate. But we know that their supply of hydrogen is finite. The first question to ask, then, is how long is a star's main-sequence lifetime?

In the latter part of the 19th century, the typical "gentleman amateur astronomer's" telescope would very likely come equipped with a device called a bifilar micrometer. This enabled the angular separation and relative orientation (the position angle or p.a.) of the components of double stars to be measured. Of particular interest were stars that were known to be physically connected, i.e., bound together and orbiting one another as a result of their mutual gravitational attraction. The ultimate reason for making this kind of observation is that it was, and still is, the only direct way to determine one of the most important numbers associated with a star – its mass.

In principle the method is "very simple"; over a period of maybe many years, you make careful measurements of the components of a double star. In time, you can work out the period "P" (in years) for the system or the total time taken for the component stars to orbit each other once. If the distance to the system is known, say from trigonometric parallax measures, then careful measurement of the angular separation of the two components can be converted into the actual maximum distance of one star from the other, "a" in astronomical units. Kepler's third law of planetary motion now tells us that the sum of the masses M_1 and M_2 (in solar masses) of the stars is given by

$$M_1 + M_2 = a^3/P^2 \qquad (1)$$

Further careful measures will enable the distances a_1 and a_2 of each star from the system's center of gravity to be determined and the ratio of the stars' masses is given by

$$M_1/M_2 = a_2/a_1 \qquad (2)$$

High school mathematics is now all that is needed to enable you to solve these two equations to calculate the mass of each star in terms of the mass of the Sun.

However, this procedure, like most things in observational astronomy, is not straightforward at all. There are many complications, for example, knowing the inclination of the binary system's orbit relative to the plane of the sky. Then there is the matter of whether both components can be clearly observed and for a long enough period of time, to determine the orbital period, and not least, of course, is knowing the actual distance to the system. However, after many years of painstaking observations, measured stellar masses began to accumulate, and when astronomers have a measured quantity for a decent-sized sample of objects, they can do great things.

When sufficient stellar masses had been determined, some of these masses could be compared with the stars' luminosities. The result was a clear and obviously tremendously important correlation between the two. Because this correlation has been arrived at by purely observational methods, it is called the *empirical mass–luminosity relation*. This relation states that the luminosity of a star L is proportional to its mass M, raised to the power of somewhere between 3.5 and 4.0.

The most important conclusions for us to draw here is that, first, if we know the mass of a star, the equations of stellar structure can be used to estimate how much of the star's mass is available as nuclear fuel. Second, by knowing the luminosity, we can estimate the expected lifetime of a star – at least while it is a main-sequence star, just as we did for the Sun in chapter *Deep Inside a Star*.

There is then, of course, the fact that if the luminosity of a star can be determined by observations (notwithstanding the difficulties mentioned in chapter *A Multitude of Magnitudes for the Colors of Starlight*), the mass–luminosity relation can itself be used to determine a star's mass.

Finally and perhaps most significantly, the fact that the luminosity is proportional to the mass raised to the power of somewhere between 3.5 and 4.0 tells us that even allowing for the fact that the nuclear burning core may differ for different stars, we can be fairly certain that not only are more massive stars more luminous but they use their nuclear fuel at a much faster rate. For example, a star of three times the mass of the Sun could shine with a luminosity 81 times greater ($3^4 = 81$). This means that it would consume nuclear fuel 81 times faster than the Sun and not just three times faster. The result is that whatever stars do as they evolve, the more massive and luminous stars will do it faster and thus have relatively short main-sequence lifetimes; fainter less massive stars, by contrast, will live life in the slow lane, but they will live for a very long time.

Finally, to give us a "feel" for actual main-sequence lifetimes of stars, we can first take that for the Sun to be 10^{10}, i.e., 10 billion years, and we can also make the approximation that the entire mass of a star (including the Sun) is available for turning into helium. We can now devise a very simple way to estimate the main-sequence lifetime of a star. Remember, that to do this for the Sun we multiply its mass $M_\odot$ by the speed of light squared (we'll call this c^2) and divide by the Sun's luminosity $L_\odot$. In other words

$$\frac{M_\odot}{L_\odot} \times c^2 = t_\odot \tag{3}$$

Here $t_\odot$ denotes the Sun's main-sequence lifetime. It follows that for any star of mass M_* and luminosity L_*, its main-sequence lifetime t_* is given by

$$\frac{M_*}{L_*} \times c^2 = t_* \tag{4}$$

The mass–luminosity relation tells us that the luminosity of the Sun and the star are each proportional to (are equal to) "a constant number" (the same number in both cases, which we can call "k"), multiplied by their respective masses, raised to some power between 3.5 and 4.0. If we use the lowest of these values, i.e., 3.5, then $L_\odot$ becomes $k \times M_\odot^{3.5}$ and L_* becomes $k \times M_\odot^{3.5}$. So now Equations (3) and (4) become

$$\frac{1}{k \times M_\odot^{2.5}} \times c^2 = t_\odot \tag{5}$$

and

$$\frac{1}{k \times M_*^{2.5}} \times c^2 = t_* \tag{6}$$

Notice how we've used the rule of indices, as described in chapter *Starlight by Numbers*, to cancel out the "M" from the numerator in each equation. For the star, we now write its mass and main-sequence lifetime as multiples or fractions of those for the Sun, so that $M_* = M \times M_\odot$ and $t_* = t \times t_\odot$ and use these in Equation (6). All we have to do now is divide Equation (6) by Equation (5), and most things will cancel out to leave us with

$$\frac{1}{M^{2.5}} = t \tag{7}$$

The mass of the star M is in solar masses, and its main-sequence lifetime is a multiple, or a fraction, of the Sun's main-sequence lifetime, which we have taken to be equal to 10^{10} years. Now we can try out this very simple equation.

First, we'll take a hot luminous star – our old friend Rigel. The mass–luminosity relation has actually been used to estimate Rigel's mass at around $50M_\odot$, so that Equation (7) tells us that t_{Rigel} equals $1/50^{2.5}$ or 5.66×10^{-5} solar lifetimes, in other words, a little over half a million years. At the other end of the main sequence lies our nearest stellar neighbor, the class M dwarf Proxima Centauri, with an estimated mass of $0.1M_\odot$ and a resulting main sequence life of $1/0.1^{2.5}$ or somewhat over 300 solar lifetimes. This is far longer than the current estimated age of the universe.

Clearly red dwarf stars are going to be around for some time. So the main-sequence lifetimes of stars range from around maybe a million years or so to what to all intents and purposes amounts to an eternity, and the deciding factor is the star's mass. The fact is, though, that stars do not just "appear" out of nowhere on the main sequence – they have to get there, and this takes us back to the beginning of the story.

The Beginning of the Story – From Atoms to Stars

Besides stars, our galaxy contains a large amount of gas and dust; the gas perhaps not surprisingly consists chiefly of hydrogen with a fair proportion of helium, and this is clearly raw material from which stars could form. We've already seen how the dust, particularly, dims and reddens starlight; it also has another role to play when stars are formed. The gas can often show itself in the form of galactic nebulae, but these nebulae glow, because either the atoms they contain ionize and recombine, as a result of high-energy radiation from hot stars within them, or they scatter the light from less luminous stars. These nebulae contain stars that have already formed, but before any stars form, a nebula is cold and dark, and this is the place where a star's life begins.

Just as a stable star exists in a state of hydrostatic equilibrium, a cloud of interstellar gas and dust is in a similar condition. The big difference is that the temperature is very low, maybe only a few Kelvin, and densities are also very low. As an example, the gas content of a cold interstellar cloud amounts to around 10^6 atoms per cubic meter. By the time this cloud gets to the stage of being a "nebula" containing stars, the density has risen to around 10^{11} atoms per cubic meter, which is still pretty rarefied when compared to the average density of Earth's atmosphere, which is about 10^{15} atoms per cubic meter.

In order to turn a cold cloud of hydrogen and helium into a nebula where stars can form, its density needs to increase, and there are several ways in which this might happen. One way is the shock wave from a supernova explosion, which can pile up and thus raise the density of intervening interstellar gas as it spreads outward. If this were the only known method of starting star formation, then there would clearly be a serious "chicken and egg" situation, i.e., how did the star that exploded as a supernova get formed?

Another process that can potentially squeeze interstellar gas on a grand scale results from galaxies colliding. However, the colliding galaxies must contain pre-existing stars, or they would not be galaxies at all but clouds of intergalactic gas. One thing that does seem clear from observations of galaxies themselves, particularly spiral galaxies, is that glowing nebulae or HII (HII standing for singly ionized hydrogen) regions, as they are often generally referred to, lie exclusively along the spiral arms of the galaxies, which thus seem to be the favored sites for stars to form. If we think of a galaxy as a revolving disk of gas initially with no stars, there will inevitably arise within the gas turbulent motion that could set up longitudinal or compression/rarefaction waves, which instead of traveling in a straight line as an everyday sound wave might, would instead move around the "galaxy's" center of gravity. In the compression zones of these revolving density waves, stars could form, but the question is, how much does the gas have to be compressed to start making a star?

This basic problem of compressing a volume of gas in order to make a star was tackled in the early years of the 20th century by the English astronomer, Sir James Jeans. Essentially the process works like this: suppose we have a large spherical cloud of gas of radius r_c which has the same temperature T and density ρ everywhere. This means that

$$\rho = \frac{M}{\frac{4\pi r_c^3}{3}} \tag{8}$$

M is the total mass of the cloud. Jeans said that under given conditions of temperature and density, the internal gas pressure of the cloud will support it against gravitational contraction, provided either its mass or its radius does not exceed certain critical values. What this means is that, if we keep the temperature and density constant but add extra layers to the cloud, thereby increasing both the total mass of the cloud and its radius, there will come a point when the cloud's internal pressure can no longer support it and it will start to contract. The critical mass here is called the *Jeans mass* and the critical value of r_c effectively measures what's called the *Jeans length*.

Both the Jeans mass and the Jeans length are inversely proportional to the square root of the cloud's density – in other words, increasing the density of the cloud means that it cannot be as big or as massive before it will start to contract. The Jeans length, which we can call l_J, is also proportional to (denoted by the symbol "$\propto$") the square root of the cloud's temperature, so that

$$l_J \propto \frac{T^{1/2}}{\rho^{1/2}} \tag{9}$$

Because the density ρ is proportional to r_c^{-3}, this means that $\rho^{1/2}$ is proportional to $r_c^{-1.5}$ and that l_J is proportional to $r_c^{1.5}$. If you're not happy with this, have a look back at chapter *Starlight by Numbers* in the section called "The Rule of Indices for all Indices."

Now imagine that we have a spherical cloud of gas whose temperature and density are such that the cloud is at its limit – i.e., its radius is equal to the Jeans length. Let's assume at this stage that whatever happens to the cloud, its temperature does not change. We shall see shortly that there is a very good reason why we can allow ourselves to make this assumption. What we want to do now is to see what happens to the cloud's Jeans length when it either expands or contracts. In order to avoid cluttering things up by including various physical constants and using the exact formula for the Jeans length (more advanced books will give you this information if necessary), we shall let the Jeans length l_J and the radius of the cloud r_c simply have a value of "1," or in other words, we don't even need to bother with the units.

Let's see first what happens to the cloud's Jeans length if the cloud expands slightly, say to a value of $r_c = 1.1$ units. Expansion means that the cloud's density falls and so the Jeans length will increase, and being proportional to $r_c^{1.5}$ it effectively becomes equal to $1.1^{1.5}$, or about 1.15 units. A bigger expansion to, say, $r_c = 1.5$ units results in an effective Jeans length of 1.84 units. Try using your calculator to input bigger values for r_c; you'll see that as r_c increases, the Jeans length increases at a faster rate, so that the cloud always stays inside its Jeans length and remains stable against gravitational contraction.

Now let's make the cloud contract to a value of $r_c = 0.9$; this results in an effective value of $l_J = 0.9^{1.5}$, or 0.85, and again $r_c = 0.5$ will give a Jeans length of 0.35. So for a contracting cloud, the Jeans length shrinks more rapidly than the cloud itself, and as we've seen this means that the cloud cannot support itself against continued gravitational contraction. Something else happens here, too; just as the Jeans length for the cloud shrinks, as the cloud contracts, so does its Jeans mass. This, it is believed, is what causes the cloud to break up into a number of fragments of varying mass (depending upon local densities within the cloud), which subsequently go on to become individual stars.

At this point, we would very likely be inclined to point out that, as our cloud began to contract, the temperature would begin to increase, and a quick look at Equation (9) tells us that an increase in temperature will increase the Jeans length of the cloud, effectively canceling out its decrease due to the increase in cloud density. However, out in the cold depths of interstellar space, where densities are very low, optical depths are also very small, save for the very central regions of the cloud. This means that any heat that is generated by the contraction can easily leak away as infrared radiation, and dust grains, as well as simple molecules, which are known to exist within interstellar clouds, actually help in this process. The result is that the initial stage of the collapse of a cloud of material to form a star occurs at essentially a constant temperature. Such a contraction is called *isothermal*, and it is in a way the very opposite of an adiabatic contraction, where no heat leaves or enters the system.

Eventually, though, as the cloud density increases, its temperature will start to rise, and the cloud's shrinking radius will start to catch up with its Jeans length. The contraction will thus slow down, but the rise in temperature means that eventually the cloud will become a significant emitter of radiation, first infrared and eventually visible radiation, whereupon it will make its debut on the Hertzsprung–Russell diagram.

At this stage the central part of our contracting cloud of gas has become a *protostar*; it will have an effective temperature of around 2,000 K and be many times the size of the Sun. It will thus resemble a red giant or supergiant and so will appear somewhere toward the upper right of the HR diagram. Exactly where it appears will depend on the protostar's mass; a more massive object does not need to contract as much as a less massive object, in order for its density to cause its Jeans length to contract to a size that is less than the size of the protostar itself. So higher mass protostars will have a larger surface area, which results in a higher luminosity. The more massive a protostar is, the further up the right-hand side of the HR diagram it will begin its *pre-main sequence evolution*.

Protostars also resemble red giants in another way – at temperatures of around 2,000–3,000 K, the opacity of material inside the protostar is relatively large, and the most efficient way for thermal radiation to reach the surface is by convection. This initially enables the contracting protostar to remain at a fairly constant effective temperature, and so as the protostar's radius and surface area decrease, the luminosity also falls. The protostar thus at first follows an almost vertical path downward on the HR diagram, which is called the *Hayashi track*, after the Japanese astronomer Chushiro Hayashi, who in the 1950s basically used the equations of stellar structure to model this stage in a star's evolution.

The lowest mass protostars contract more slowly, and so the compression of the gas in the protostar is more nearly isothermal. Provided the protostar's mass is sufficient (the minimum is estimated to be around $0.1M_\odot$), temperatures at the center will reach around 10^7 K, enabling the proton–proton chain reaction to get going. This low-mass protostar has now become a true star at the lower end of the main sequence – in other words, an M dwarf.

Also working on pre-main sequence evolution at the same time as Hayashi was the American astronomer Louis Henyey and his colleagues, who showed that for more massive protostars (i.e., those more massive than around $0.5M_\odot$), the Hayashi track stage is relatively short lived (and the more massive the protostar the shorter this stage becomes), and it is followed by a more or less horizontal track (actually called the *Henyey track*) from right to left on the HR diagram. In other words, the protostar's effective temperature is increasing, but its luminosity is remaining more or less constant. More massive protostars contract more rapidly, and the resulting compression is in effect more adiabatic, which means that there isn't sufficient time for thermal radiation to escape at a rate that would maintain a constant temperature.

The resulting rise in effective temperature would normally increase the protostar's luminosity, but it is still contracting, and the decreasing surface area effectively counteracts this. Hence the Henyey track is more or less horizontal.

Another thing that happens with more massive protostars is that the higher temperatures generated inside mean that the opacity of the protostellar material falls (remember Kramers' Law from chapter *Deep Inside a Star*), and so the envelopes of more massive stars become radiative rather than convective. The relatively rapid temperature rise for more massive protostars also means that the central core temperature more quickly reaches the proton–proton chain temperature of around 10^7 K.

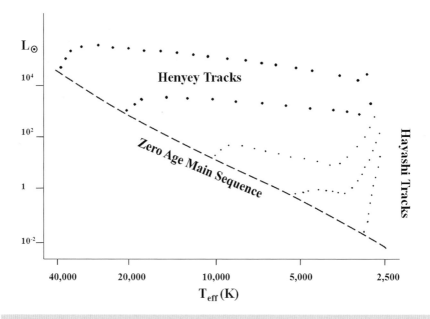

Figure 1. Broadly speaking, the evolutionary tracks across the HR diagram for contracting protostars divide into two parts. The Hayashi track is more or less vertical and more extensive for lower mass stars, whereas the roughly horizontal Henyey track dominates the pre-main sequence evolution of more massive stars. The time spent on these tracks ranges from around 10^4 years for the most massive protostars to around 10^7 years for those of lower mass.

The result is that the more massive a protostar, the shorter the time from its first appearance on the HR diagram to arriving on the main sequence. The core temperatures of the most massive stars will continue to rise still further, which then favors the CNO cycle. The end points of the Hayashi/Henyey tracks form a curve, which follows the main sequence on the HR diagram. This curve is referred to as the *Zero-Age Main Sequence,* or ZAMS for short. Figure1 shows schematic pre-main sequence tracks across the HR diagram.

So now we're back at the middle of the story, but before we move on to the next part of the story, it's important to point out that the main sequence is not a "line" on the HR diagram but what amounts to a fairly broad band of stars. The fact is that even when a star has reached its place on the ZAMS, its core is still contracting, albeit fairly slowly, in order to achieve hydrostatic equilibrium. This raises the core temperature enough to expand the star's envelope to some degree, and indeed it has been estimated that during the course of its main-sequence lifetime, the Sun will have increased its size by a factor of about two. All this time, of course, the hydrogen in the stellar core is becoming more depleted, so that eventually a loss of thermal equilibrium results in a much more serious disturbance to the star's hydrostatic equilibrium situation. This results in the star having to essentially "reinvent" itself, and in doing so, it will leave the main sequence and pursue what to most astronomers is by far the most interesting part of its life.

The End of the Story – Life After the Main Sequence

When most of the hydrogen in the core of a star has turned into helium, the core will for a time be inactive as far as nuclear reactions are concerned and will contract under its own gravity. This will happen whether the star is massive or not so massive. However, the consequences of this core contraction will be somewhat different for stars of relatively low mass, say, less than about $2M_\odot$, compared to that for higher mass stars; ultimately, though, it involves a bit of bizarre physics for both groups of stars.

In a lower mass star, the inactive helium core will initially have a temperature of around 15 million Kelvin, and as the core contracts its temperature will start to rise. As if jumping on the band wagon, the layer of hydrogen immediately surrounding the core, which hitherto was not hot enough for hydrogen fusion, will also contract and heat up to the point where hydrogen fusion can begin. This now active shell of hydrogen in turn heats and expands the main part of the star's envelope. As it expands, it cools and turns our low mass star into a red giant.

In the case of a star of $1M_\odot$, this process is estimated to take around 1 billion years, with stars of slightly higher or lower mass taking correspondingly shorter or longer times, respectively. If we could speed the process up, we would see the star move upward and somewhat to the right in the HR diagram, resulting from a drop in effective temperature as the envelope expanded, but a rise in luminosity as the area of the star's photosphere increased in size. The star would be on the *red giant branch*, or RGB, of the HR diagram.

Meanwhile, back in the core, strange things are beginning to happen. The contraction of the core under its own gravity ultimately forces the two electrons in each helium atom into the lowest energy levels. In very basic terms, the helium atoms become physically smaller in size, as they squeeze together, while still maintaining their identity as actual atoms. While this is going on, the gas pressure in the core is increasing, as one would expect from the normal kinetic theory of gases. Then, a fundamental law of quantum mechanics called the Pauli Exclusion Principle (named after the Austrian physicist Wolfgang Pauli) comes into operation. This law states that, within an atom, only one electron can occupy a given energy level at any one time, so that while an atom has one or more electrons in its higher energy levels, there is always "somewhere," i.e., some vacant lower energy levels, that they can go to if need be. With electrons filling all the lowest energy levels from the "bottom up," there is nowhere for the electrons to go, and any material which is in this condition is called *degenerate*. Degenerate electrons in this particular case resist being compressed any further by a force, which is referred to as "electron degeneracy pressure," and this is the maximum pressure that the stellar core is able at this point to exert.

From this stage any energy that is expended by gravity will barely compress the star's core any further, and indeed extra gravity does "arrive" in the form of helium that has been created in the hydrogen burning shell and which adds mass to the core itself. The effect, though, is to make the helium atoms themselves kind of "slip and slide" around one another at increasingly greater speeds; in other words, the gravitational energy is now used solely in raising the temperature of the core itself, until it eventually reaches a "magic" figure of around 10^8K. At this temperature the non-degenerate helium nuclei basically "forget" about the degenerate electrons and fuse together in a (for this star) brand new thermonuclear reaction called the *triple-alpha process*.

The end product of this process is carbon, and indeed some of the resulting carbon atoms can fuse with another helium nucleus to make oxygen. Meanwhile, the release of this new source of thermonuclear energy in turn releases the electrons from their degenerate grip. Because of the forgoing degeneracy situation and the buildup of the core temperature to a point that is aptly described as "critical," the initiation of helium fusion in lower mass stars is relatively rapid – more in the form of an explosion deep inside the star, which is referred to as the *helium flash*.

The removal of electron degeneracy in the core by the helium flash means that the core behaves once more according to the good old rules of the kinetic theory of gases; it expands and cools, slowing down the helium fusion reactions. This core cooling initially makes the star as a whole contract somewhat, but then as hydrostatic equilibrium is re-established, energy transport through the envelope (this time by convection) halts the contraction, while at the same time raising the star's effective temperature. The star thus moves more or less horizontally to the left across the HR diagram, along what is known as the *horizontal branch*.

There is one more phase in the post-main sequence story of our low mass star, before it reaches its "endgame." In time the fusion of helium will cease (estimated to occur after about 100 million years for the Sun), and there will remain an inactive carbon/oxygen core. This will in turn contract and become degenerate again, but with a relatively low mass core this contraction will never generate a high enough temperature for carbon atoms to fuse.

However, something else does happen. The hydrogen burning shell has now given rise to a shell of helium surrounding the core, which as a result of the core's contraction will itself contract, heat up, and undergo a second so-called "helium shell flash." This flash is not as violent as the core helium flash, because the density here is too low for the gas to become degenerate, and indeed it can occur in several stages called thermal pulses, which give rise to instability in the star's envelope. This will expand our humble low mass star's outer layers even further to the extent that it will move for a time right up to the top right corner on the HR diagram, on what is called the *asymptotic giant branch*, or AGB.

The thermal pulses taking place within an AGB star are believed to contribute to these stars' pulsational instability, resulting in most of them being long period or semi-regular variable stars. Probably the most famous AGB star is Mira.

Finally, the vast and relatively low-density envelopes of these stars (a $1M_\odot$ AGB star is estimated to have a radius of around 1.5 A.U.) mean that much of this material is only held loosely by gravity to the star's core. The result is that thermal pulses help to drive material away in the form of a low velocity stellar wind. As mentioned in chapter *Deep Inside a Star*, convection currents dredge up material, such as carbon and oxygen, that has been synthesized in the helium burning shell.

The final result of this process is to drive most of the star's envelope away from the core in the form of a large low density "bubble," which is photoionized by high-energy photons from the still very hot (around 10^5 K) core – a planetary nebula of the kind we all know and love.

In this situation, our low mass star has finally, albeit in "death," won its constant battle with gravity to maintain hydrostatic equilibrium. This is achieved by the electron degeneracy pressure of the remnant core, which will maintain hydrostatic equilibrium indefinitely as it cools. This core is now a white dwarf star – it's hot but small (approximately the same size as Earth) and hence of low luminosity. This puts its final resting place near the bottom left corner of the HR diagram, somewhere way below the main sequence.

Back in chapter *Deep Inside a Star*, we pointed out that much of a star's mass is concentrated in the core, and this means that higher mass stars have more massive cores. This in turn means that much higher temperatures can be generated within them, when they contract. The first effect of this is that the contracting helium core of a massive star does not have to reach the stage of degeneracy before helium fusion begins, and as before, a shell of fusing hydrogen takes our more massive star on what this time turns out to be its first "leg" along the giant branch.

For a more massive, and initially much hotter, star, the relative drop in effective temperature is much greater as it evolves to a red giant, and this offsets the effect of an expanding photosphere. The result is that more massive stars follow a more horizontal RGB across the HR diagram. Although there is no helium flash, it has been suggested that for what are generally referred to as "intermediate mass stars" (stars of mass from around $2.5M_\odot$ to $5M_\odot$), there could be a "carbon flash," as their more massive and this time degenerate carbon cores initiate carbon fusion. This happens at a temperature of around 500 million Kelvin, the fusion products being neon, sodium, and magnesium.

What is particularly unclear is whether such a carbon flash could actually blow the star apart, mainly because it is not known with any certainty how many neutrinos are produced in such high temperature fusion reactions. In sufficient numbers, they could in theory remove a significant proportion of the energy produced in a carbon flash, and, as a result, cool the stellar core; but this one remains firmly in the hands of the theorists.

Beyond this, thermonuclear reactions for increasingly massive stars become much more involved and, to be honest, are probably still not fully understood in all their details. There are a few general points, though, that can effectively "pull together" this phase in a massive star's life. First, the onset of a new form of fusion process will initially cause the stellar core to expand and cool. As with lower mass stars, this will result in the star as a whole contracting for a time. This will regulate the fusion reaction rate and re-establish hydrostatic equilibrium. Thermal energy transport will then in turn cause the star to move to the left, along the horizontal branch, as its effective temperature rises again. For the most massive stars, this process can happen several times and results in them pursuing a kind of "to and fro" track across the HR diagram.

Second, the actual "yield," or the amount of energy that is released by the fusion of progressively heavier atomic nuclei, is in turn progressively lower, though the actual reactions themselves are capable of synthesizing pretty well all of the chemical elements in the periodic table up to and including iron. A very important consequence of the post-main sequence tracks of stars across the HR diagram (this includes lower mass stars) is that at some point in time they cross, maybe more than once, the Cepheid instability strip. This, as we have seen, makes them prone to sustained pulsation within their envelopes and tells us, of course, that Cepheid and other related variable stars are just "passing through" on their evolutionary journey across the HR diagram.

Finally, just as a hydrogen fusing shell developed around the helium core of our low mass star, here a whole series of fusion shells progressively develops, the outermost consisting of hydrogen, which surrounds a helium shell, which in turn surrounds a carbon shell, etc. The actual number of these fusion shells is determined by the mass of the star itself, but even for the most massive stars, the innermost shell ultimately consists of silicon, which is fusing to iron, which adds to the mass of the star's iron core. The formation of this core marks the endgame for the most massive stars. The post-main sequence tracks for stars across the HR diagram are shown schematically in Fig. 2.

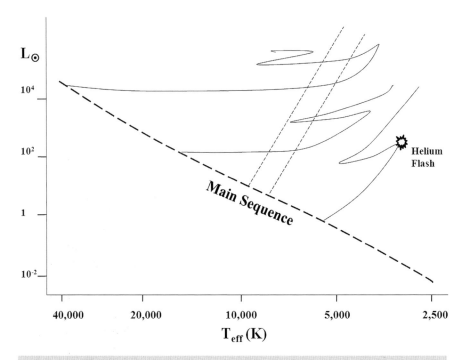

Figure 2. Schematic post-main sequence evolutionary tracks for low, intermediate, and high mass stars. The parallel *dashed lines* show the approximate location of the instability strip and indicate the fact that higher mass stars cross this zone several times.

The "fusion game" ends with the production of an iron core. Fusing iron to make heavier elements actually takes energy in, rather than producing it, and so when the last silicon atoms fuse to make iron atoms, the core contracts for the last time. The increasing density will, as with the star's low mass cousins, render the core degenerate, but this time there is no release mechanism in the form of yet another fusion process that can free the degenerate electrons. Instead, the extremely high temperature within a high mass stellar core (of the order of 10^{10} K) starts to break the iron nuclei apart again, in a process known as *photodisintegration*.

The energy contained within the core, which has hitherto been used to maintain thermal equilibrium and hydrostatic equilibrium throughout the star, is now being used by the core to effectively destroy itself. This actually releases the electrons, but the consequence is truly catastrophic. The products of photodisintegration of iron atoms are protons, electrons, and helium nuclei, and under these conditions, the electrons – all the electrons – readily combine with the protons to make neutrons. What's more, because the electron degeneracy pressure has gone, the core collapses, which in turn squeezes the neutrons together, so that they become degenerate. The resulting neutron degeneracy pressure prevents the core from further collapse – but maybe only for the moment.

The loss of thermal equilibrium results in the loss of hydrostatic equilibrium throughout the star. The result is that the entire outer layers of the star implode onto the core of degenerate neutrons – and bounce back to produce a Type II supernova. This, as every astronomer knows, is an almighty explosion, and as one amateur astronomer said "something that we really are overdue for in our galaxy."

What of the stellar core, though? We have a situation where the irresistible force of gravity meets with the immovable object in the form of a pile of degenerate neutrons – a *neutron star* – and provided the mass of this dead stellar core isn't too high, neutron degeneracy pressure maintains hydrostatic equilibrium indefinitely. Once again the star wins, this time having truly gone out in a blaze of glory.

Neutrons are, of course, electrically neutral particles, but the outer layers of a neutron star are in fact a plasma, consisting of atomic nuclei and electrons. What's more, neutron stars spin very fast, due to a basic principle of physics called *the principle of conservation of angular momentum*. This principle was "there" back at the start of a star's life. It made the interstellar cloud begin to rotate as it contracted. It works very simply. Any object, or particle for that matter, that is moving possesses, as we have seen, kinetic energy as a result of its motion. The particle also possesses another quantity, intimately connected to its kinetic energy, called its *momentum*, or more strictly its *linear momentum*, and this is simply equal to the particle's mass "m" multiplied by its velocity "v," or $m \times v$. Linear momentum is in a sense a measure of how much force you would have to apply to stop a body from moving. It would take a lot of force to stop a slow-moving massive object such as a locomotive, but it would also take a lot of force to stop a fast-moving low mass object such as a bullet.

If, instead of moving in a straight line, our particle is effectively moving around some fixed point, which is at a distance "r," then its linear momentum, which is still equal to $m \times v$, gives the particle a kind of "lever effect" around the fixed point. The greater the particle's distance from the point, the greater this lever effect is. This is called the particle's *angular momentum*, and it is equal to $m \times v \times r$. The principle of conservation of angular momentum says that if the particle's distance from the point changes, its angular momentum must nonetheless remain constant. This means that if the distance decreases, the particle must speed up to compensate and vice versa. A particle within a contracting stellar core must maintain its angular momentum around the core's axis of rotation, and so it, and all its buddies, will speed up; in other words, the stellar remnant spins faster.

The rapid rotation of the plasma-rich surface layers of a neutron star, combined with its extremely small size, result in an immensely intense magnetic field. Charged particles suffer severe acceleration in this field and emit electromagnetic radiation, often of very short wavelength, predominantly along two narrow beams that are directed along the axis of the star's magnetic field. The rapid rotation of the star itself can result in one of these beams being observed as a rapid succession of light pulses – known as a *pulsar* – of which the most famous example is the one that lies at the heart of the Crab Nebula supernova remnant in the constellation of Taurus. Over recent years more exotic objects, such as quark stars (quarks being the fundamental building block of particles such as protons and neutrons) and even "preon stars," (preons, in turn, being the hypothetical building blocks of quarks) have been proposed, but with sufficient mass remaining in a stellar core, gravity ultimately does appear to win.

The English physicist Stephan Hawking showed us that, over a period of time, black holes could lose mass and "evaporate." The fact is that it wasn't until a few decades ago that most astronomers were generally very skeptical about the very existence of black holes. What changed things was not so much the idea that massive stars at the end of their lives could turn into black holes but that they offered a very satisfactory explanation of the enormous luminosity of quasars. By contrast, physicists had grown accustomed to the idea of black holes in the late 1940s, when J. R. Oppenheimer used the general theory of relativity to investigate the gravitational collapse of massive objects such as very massive stars. Physicists were generally "okay" with the idea that a star could collapse to the point where the escape velocity from its surface exceeded the speed of light and thus form an *event horizon*, inside of which nothing could escape. However, the idea that the collapsing object could continue right down to zero size and infinite density, in other words a *space-time singularity*, as general relativity seemed to predict, was seen as highly improbable.

Thus, it may be said that Hawking's finest achievement was theoretical work carried out in the late 1960s and early 1970s, partly in collaboration with the Oxford mathematician Roger Penrose, on what are now called the *singularity theorems*. These basically showed that, provided sufficient mass can be compressed to the extent that an event horizon does form, then the inevitable consequence is a true space-time singularity. The final bottom line here, where gravity seems to have won, is that the existence of black holes implies the existence of singularities, which in turn implies that there are regions of the universe where our most fundamental ideas of physics run out – including general relativity. So, in the end, it may be that gravity as we know it hasn't won at all, but instead it is as-yet unknown physics that has finished the job.

Stellar Evolution in Action

Probably the most important underlying feature of the evolution of stars is that more massive stars do it faster – and this means in all stages of their evolution. The fact that our galaxy contains stars in all stages of evolution is clearly demonstrated by their various locations on the HR diagram. The galaxy also contains stars of all ages, because as we shall now see, if all the stars in the galaxy were of the same age, the HR diagram would look very different.

There are places in the galaxy, however, where all the stars *are* of the same age, and this is in star clusters – open or galactic clusters, such as the Pleiades and globular clusters such as M13. The starting point for all star clusters is a contracting cloud of interstellar gas that, as indicated above, will fragment as the Jeans mass in different parts of the cloud decreases, at a rate that depends on the local density. These fragments will go on to form a cluster of stars, in which the most massive fragments will become the most massive and the fastest evolving stars. Let's see, then, what the HR diagram for an evolving star cluster would look like.

The cluster will make its debut on the HR diagram as a growing group of low temperature objects on the right-hand side. These objects are following their respective Hayashi tracks. The most massive of these objects will then move toward the left, as they pursue their Henyey tracks, and these, the most luminous hot stars, will be the first stars to reach what will be the very top of the ZAMS. The lower parts of the ZAMS will start to fill, as the less massive protostars begin fusing hydrogen in their cores and achieve hydrostatic equilibrium, but by the time the lower mass protostars are reaching the ZAMS line, the most massive stars have already evolved away from the main sequence to become red giants.

As progressively lower mass stars evolve up the giant branch, the main sequence basically "peels away," at what is referred to as the "turn-off point" from the top down, and indeed, the degree to which this has progressed gives astronomers a real measure of the star cluster's age. Meanwhile, the most massive stars have now evolved along the horizontal branch and onto the AGB, prior to perhaps exploding as supernovae, and in time, the AGB will fill with less massive stars on their way to becoming planetary nebulae.

The main difference, then, between a general HR diagram and that for a star cluster is that the cluster diagram will have a piece of the upper main sequence missing. In addition, a well-populated horizontal branch and AGB tells astronomers that they are dealing with an old star cluster. In this way, astronomers know that globular clusters are generally very old indeed – maybe as much as 10 billion years, as compared with galactic clusters, which in some cases are of the order of only a few million years old. There are, of course, exceptions, such as the very old galactic cluster M67 in Cancer and the relatively young globular cluster M71 in Sagitta.

Finally, a tremendous bonus for both professional and amateur astronomers who wish to investigate star clusters is the fact that all the stars in a cluster are effectively at the same distance from us, and so the stars' luminosities scale directly with their apparent, as well as their absolute, magnitudes. This means that when producing an observational HR diagram for a cluster, it is only necessary to plot the stars' V magnitudes against their B – V color indices. Figure 3 shows schematic HR diagrams for young, intermediate, and old star clusters

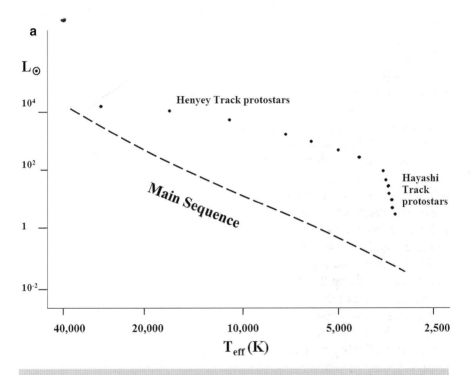

Figure 3. Here we see three stages in the evolution of a star cluster. As the cluster forms, lower mass stars move on Hayashi tracks toward the ZAMS, while the more massive stars pursue a more rapid course along Henyey tracks and reach the ZAMS first, as shown in a. In a somewhat older cluster, intermediate mass stars have reached the ZAMS, whereas low mass stars have yet to reach it. Meanwhile, the highest mass stars are already in their post-main sequence phase, as shown in b. For an old cluster, such as a globular cluster, all stars initially on the upper part of the main sequence

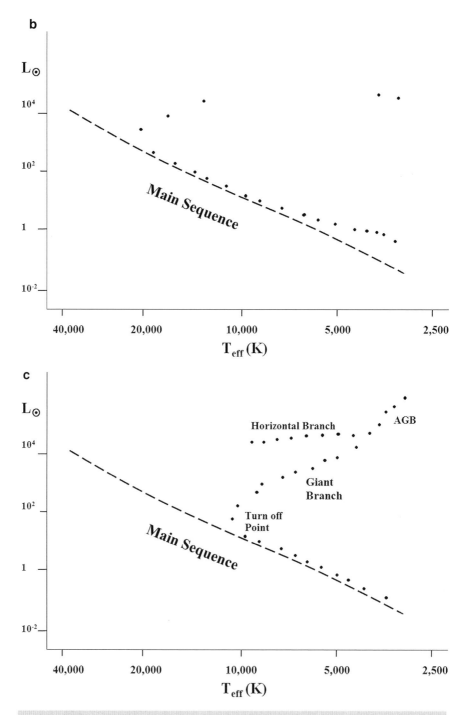

Figure 3. (continued) have evolved to produce well-populated horizontal branch and AGB regions as shown in c.

On a final note, the globular cluster M71 mentioned above was initially thought to be a fairly dense galactic cluster, until more detailed observations showed that its stars were more like those of a globular cluster than a galactic cluster. The difference here is not just one of age but of actual chemical composition.

Old Stars, Young Stars, and New Stars from Old

The first stars that ever formed consisted of hydrogen and helium together with a trace of lithium and beryllium, because these were the only elements that were present after the Big Bang. We've now seen that stars spend their lives synthesizing heavier chemical elements by nuclear fusion in their cores; more helium is made, followed by the likes of carbon, oxygen, etc., right up to and including iron. What's more, it is believed that other elements in this part of the periodic table are produced in "side reactions" that take place alongside the main fusion process. What we see here is the fact that the stars are actually the universe's chemical factories; what we've also seen is that elements further up the periodic table cannot be synthesized in fusion reactions. These elements are "here," though, right up to element number 92, which is uranium.

It's now known that in the more massive AGB stars, heavier and rarer elements can form within some of the fusion shells that these stars contain, and the general name for the processes that occur here is the "*S-process*," which stands for slow neutron capture. Any chemical element is "defined" by the number of protons in the nuclei of its atoms, and the nuclei of all elements have a slightly variable number of neutrons. The actual number of neutrons "defines" an isotope of a particular element, and while some isotopes are stable, others are not. One of these extra neutrons can split up or "decay" by emitting an electron – in a process called β or "beta" decay – because in the years following the discovery of radioactivity, electrons emitted by atomic nuclei were called "beta rays." This then leaves behind an extra proton in the nucleus, which means that our element is no longer the same element but has in fact become the next element up in the periodic table.

In the fusion shells of AGB stars, "free" neutrons exist and take part in the main fusion reaction processes, and each neutron capture/decay event leads to a nucleus of a heavier element. With the S-process it is a step by step process of element synthesis, because the quantity of neutrons that are present in the fusion shells of AGB stars is relatively low, so that β decay takes place before an atomic nucleus can grab itself one or more further neutrons. Even so, in time, fairly heavy elements can be produced in this way, and as we have seen, convection processes within these stars can dredge these elements up into the stars' surface layers, where they show themselves in the stars' spectra (indeed the first hard evidence for the S-process came in 1952 when the American astronomer Paul Merril discovered the element technetium in the spectra of some red giants) and may subsequently be sent off into interstellar space, as part of a stellar wind.

The collapse of a supernova provides a very high abundance of free neutrons, which can attach themselves particularly to the iron nuclei in the core. What's more, in this situation, several neutrons may be captured before any β decay takes place. This process of "rapid neutron capture," known as the R-process, can lead especially to the synthesis of the heaviest radioactive elements, such as uranium. The explosive character of a supernova means, of course, that much of this heavy element material is expelled as part of a supernova remnant.

The bottom line here is that stars enrich the interstellar medium with heavier elements, which means that younger stars will inevitably contain a higher proportion of these heavier elements than the old ones. This higher heavy element content will reveal itself in the presence of more and stronger absorption lines, due to heavier elements in the spectra of younger stars. This kind of distinction in stellar spectra was first recognized in the mid-1940s by the German-American astronomer Walter Baade, who defined the terms "population I" and "population II" when referring to younger and older stars, respectively. Since that time, the original division has developed more into an evolutionary scale, with the terms "extreme, intermediate, and old" being applied to the youngest through to the oldest population I stars and "young, intermediate, and extreme" being applied to the youngest through to the oldest population II stars. Our Sun is considered to be an old population I star.

The Legacy of Starlight

Stars end their lives as cooling white dwarfs, neutron stars, or black holes, but while they shine, they synthesize virtually all of the chemical elements that feed into the interstellar medium. This material goes on to make new generations of stars, which themselves contain a greater abundance of heavier elements. The formation of stars in an environment that is relatively rich in heavy elements results in "heavy element planets," such as Earth, forming as part of the process. This, then, is the wonderful legacy left to us by the stars; in our part of the universe, the ultimate legacy of the river of starlight – is us!

Conclusion

This, then, has been our basic introduction to stellar astrophysics. As mentioned at the beginning, it is an enormous subject with many specialist areas that themselves would in some cases require a whole other book to deal with properly. Even so, if what you have read has taught you something about stars that you didn't know before and maybe even opened your eyes to an area of astronomy that has hitherto been unknown, then that is certainly reward enough for this author. Most of all, if it has in some way given a little more meaning to your observations, then it has done its job.

Clear skies and good observing!

Key Points

- The luminosity of a star is proportional to its mass raised to a power that lies in the range of 3.5–4; this is called the mass–luminosity relation. It can be used, for example, to estimate the main-sequence lifetime of a star.

- The more massive a star is, the more rapidly it proceeds through all stages of its evolution.

- If temperature and density conditions within an interstellar cloud are such that the linear dimensions of the cloud exceed its Jeans length, or its total mass exceeds its Jeans mass, then the cloud will contract under its own gravity.

- It is believed that local density variations in a contracting cloud will result in the local Jeans mass decreasing at different rates in different parts of the cloud, leading to fragmentation into units that will ultimately form stars of different masses.

- A contracting cloud fragment will heat up and eventually appear on the right-hand side of the HR diagram, whereupon it will follow an initially vertical Hayashi track as it contracts isothermally.

- More massive cloud fragments contract more rapidly, i.e., more adiabatically, and follow a more horizontal Henyey track to the left of the HR diagram.

- When the cloud core reaches a temperature of around 106 K, thermonuclear fusion of hydrogen begins, and the star has arrived at the zero-age main sequence, or ZAMS.

- Lower mass main sequence stars, in time, switch from fusing hydrogen to helium, in favor of fusing helium to carbon and some oxygen in the triple-alpha process.

- Low mass stars end their lives on the asymptotic giant branch, or AGB, of the HR diagram, where they in time expel their outer envelope to become a white dwarf surrounded by a planetary nebula.

- More massive stars fuse increasingly heavy elements, initially in their cores, and later in a series of shells surrounding the core. They end their lives by their outer layers imploding onto an inert iron core, the "rebound" manifesting itself as a supernova.

- Succeeding generations of stars enrich the interstellar medium with heavier elements, most of which are synthesized by either the slow or rapid neutron capture processes.

- Heavier elements made in stars make planets such as Earth and living creatures like us.

Appendix 1: The Greek Alphabet

This is usually given in most books on observational astronomy, but we've used so many Greek letters in our various equations that we thought it would be helpful to provide a list of all the lower case letters here.

α	Alpha	ν	Nu
β	Beta	ξ	Xi
γ	Gamma	o	Omicron
δ	Delta	π	Pi
ε	Epsilon	ρ	Rho
ζ	Zeta	σ	Sigma
η	Eta	τ	Tau
θ	Theta	υ	Upsilon
ι	Iota	φ	Phi
κ	Kappa	χ	Chi
λ	Lambda	ψ	Psi
μ	Mu	ω	omega

K. Robinson, *Starlight*, Patrick Moore's Practical Astronomy Series,
DOI 10.1007/978-1-4419-0708-0, © Springer Science+Business Media, LLC 2009

Appendix 2: Astronomical and Physical Units and Constants

The word "units" means the basic quantities in which things are measured; for example, distance may be measured in miles or kilometers. Astronomers can be notorious for using units that often seem a bit bizarre, such as giving the diameters of stars in centimeters. By contrast, physicists have over the years standardized things a lot, and they now very widely use the so-called MKS (for meter – kilogram – second) system, which is also referred to as the "SI" system, which in French stands for "Le Système International d'Unités."

The fact is that many quantities are actually a combination of units of mass, length, and time (hence "meter, kilogram, second") and maybe also temperature. A good example is energy, which in the SI system is of course measured in joules. Energy is equivalent to the work that is expended or done by some system, and work is defined as force multiplied by distance. According to Newton's second law of motion, force is equal to mass multiplied by acceleration; acceleration is the rate at which velocity changes and so is equal to velocity divided by time. Finally, velocity is equal to distance divided by time. So, denoting distance by "L," mass by "M," and time by "T," we have

Velocity $= L/T$
Acceleration $= L/T/T = L/T^2 = LT^{-2}$
Force $= M \times$ acceleration $= MLT^{-2}$
Energy $=$ force $\times$ distance $= ML^2T^{-2}$

The combination ML^2T^{-2} is referred to as the *dimensions* of energy (nothing to do with dimensions of space and time), and indeed all physical quantities have their respective dimensions. In an equation, dimensions will multiply together and cancel out just like numbers do; a good example here is Newton's famous inverse-square law for the gravitational force "F" between two bodies each of mass "M" and separated by a distance "L." This is given by $F = G \times M^2/L^2$, and G here is the universal constant of gravitation. We have seen that force has the dimensions of MLT^{-2} and so the dimensions of G are given by $MLT^{-2} \times L^2/M^2$, which is the same as $L^3M^{-1}T^{-2}$.

An important general rule and a very valuable check in any equation involving physical quantities is that the dimensions of each side of the equation should be the same.

Important Units

In addition to the fundamental units of kilogram (kg), meter (m), and second (s), there are several basic units that, besides being important in themselves, can also serve to define the units of other quantities. These are

- Force – the *Newton* – "N:" defined as the force that will give a mass of 1 kg and an acceleration of 1 ms^{-2}.

- Energy – the *joule* – "J:" defined as a force of 1 N moving through a distance of 1 m.

- Power – the *watt* – "W:" defined as energy expended at a rate of 1 Js^{-1}.

- Temperature – the *Kelvin* – "K:" defined as one degree on the Kelvin temperature scale, such that 0° Celsius equals 273 K.

- Frequency – the *Hertz* – "Hz:" defined as one oscillation per second.

Notice here that it is very common practice to write, for example, "s^{-1}" instead of "per second," or "K^{-4}" instead of "per Kelvin to the power four." This saves an awful lot of time.

Constants

Although all constant have, as we saw above, their own respective dimensions, it is nonetheless standard practice to give them in their own respective units. So for example the units of the constant of gravitation are Nm^2kg^{-2}, which *do*, of course, reduce to the dimensions given above. A good exercise is to work out what the dimensions are for the various constants given below. The actual value for a constant will more often than not be a number that is given in scientific notation.

Physical Constants

The speed of light – $c =$	$2.9979 \times 10^5 \text{ m s}^{-1}$
Gravitational constant – $G =$	$6.673 \times 10^{-11} \text{ N m}^2 \text{ kg}^{-2}$
Planck's constant – $h =$	$6.626 \times 10^{-34} \text{ J s}$
Boltzmann constant – $k =$	$1.381 \times 10^{-23} \text{ J K}^{-1}$
Stefan's constant – $\sigma =$	$5.670 \times 10^{-8} \text{ W m}^{-2} \text{ K}^{-4}$

Astronomical Constants

Astronomical unit – A.U. $=$	$1.496 \times 10^{11} \text{ m}$
Parsec – pc $=$	$3.086 \times 10^{16} \text{ m}$
Light year – ly $=$	$9.461 \times 10^{15} \text{ m}$
1 pc $= 3.2616$ ly	
Solar mass – $M_\odot =$	$1.989 \times 10^{30} \text{ kg}$
Solar radius – $R_\odot =$	$6.96 \times 10^8 \text{ m}$
Solar luminosity – $L_\odot =$	$3.826 \times 10^{26} \text{ W}$

Appendix 3: The Doppler Effect

There are probably few astronomy books these days that have nothing to say about the famous Doppler effect, named incidentally after the early 19th-century Austrian physicist Christian Johann Doppler. Doppler successfully explained the phenomenon of wave motion that affects the wavelength and also consequently the frequency of waves that are emitted either by a source whose motion has a component directed toward or away from the observer or are received by an observer who is moving relative to the source. This means that light from a star whose motion is partially or wholly directed toward or away from us is slightly bluer or redder, i.e., of slightly shorter or longer wavelength, respectively, as a result.

The most sensitive and accurate way to measure such changes in wavelength is by using the shifts or wavelength changes in the absorption lines in the spectra of stars. The change in wavelength "$\Delta\lambda$" (pronounced "delta lambda") for an absorption line whose laboratory or "rest" wavelength is λ_0 is given very simply by

$$\Delta\lambda = \lambda_0 \times \tfrac{v}{c} \tag{1}$$

Here, v is the relative velocity of the source and the observer, and c is the speed of light in the *same units* as that of v. If λ_0 is in angstroms, then $\Delta\lambda$ is also in angstroms, i.e., *the same units* once again.

Finally, note that the standard convention is for relative motion directed away from the observer to have a positive value for v and vice versa. Thus a positive value for $\Delta\lambda$ denotes a red shift; likewise, a negative value denotes a blue shift. Equation (1) is perfectly okay for most astronomical situations where v is small compared to the speed of light, so that there is no need to use the more complicated relativistic Doppler formula that, needless to say, came some time after Professor Doppler.

Index

Printed in the United States of America